GETTING THE BEST RETURN

THE POWER OF STATISTICS

GETTING THE BEST RETURN

THE POWER OF STATISTICS

Trond Brekke

LIFESUCCESS PUBLISHING, LLC
8900 E. Pinnacle Peak Road, Suite D240
Scottsdale, AZ 85255

Telephone:	800.473.7134
Fax:	480.661.1014
E-mail:	admin@lifesuccesspublishing.com
ISBN:	978-1-59930-212-6

Cover:	Daniela Savone, LifeSuccess Publishing, LLC
Text:	Lloyd Arbour, LifeSuccess Publishing, LLC
Edit:	LifeSuccess Publishing, LLC; Publication Services, Inc.

Printed in Canada.

DEDICATION

To my son, Bob, my greatest inspiration.

CONTENTS

FOREWORD

Statistics inform us and help us make critical decisions on a daily basis. Polls, surveys, and factual research help communities decide what political party to support, how our health-care system is faring, and how well our kids are being educated in school. Have you ever stopped to think about how statistics can help you improve your own life?

For some people, success just seems natural. However, successful people have one thing in common. They know how to extract and analyze information from their own life. In a way, it's like they're gathering research on facts and personal statistics and then measuring the probability of choices; they do this so that they can make informed decisions that will support their goals and dreams.

Trond Brekke is one person who knows how to use personal statistics to his advantage. *Getting the Best Return* is a gold mine of information on how you can start analyzing your own life so that you can get a better return. The point of *Getting the Best Return* is to teach you how to reach your potential through making conscious choices and acting upon them. Trond will show you how powerful you truly are and how easy it is to change your life forever and for the good.

No matter what your life has been like in the past and what systems you have tried before, Trond's system is one that works. Leave behind the mindsets that are not working for you and begin looking forward to a healthy life of good relationships, great jobs, and worry-free finances. This book will teach you how to discover the power that is within you; you can change your life simply by taking charge of your decisions and actions. Let Trond teach you how to have the life you have always desired – a life that you have the power to create yourself.

—Gerry Robert,
bestselling author of *The Millionaire Mindset*

ACKNOWLEDGEMENTS

So many people have been involved in my life that it is almost dangerous to start mentioning names. I will give Bob Proctor and Gerry Robert most of the blame for me writing this book. I will also mention my son, Bob. The book is partly written so he can have it as a guideline when he grows up, hopefully proud of his author father. Also a big thank you to all the others that have influenced my life.

INTRODUCTION

Many of us have thought that making significant changes in the quality of our lives would be a great idea. However, few of us know how to do this effectively. What we lack is a plan, a way to effect change in our lives. How are we to know where to begin? The changes we want to make may be in the way we think, in our personal relationships, in our financial situation, or in our professional life. We may even want to make serious changes in our career path, our family, or our general perspective on life.

The power of statistics helps you understand your place in the world and make decisions that will lead you toward your goals. With this program, you will learn how to make conscious, positive decisions, how to keep from being misled by your habits and the mindset that you are stuck in, and how to clearly change your life forever and for the better.

If you have ever thought about making a change in your life, now is the time to actually do so. Many of us allow our fears and doubts to take over our lives, and we become unable to act. When we cannot act upon our goals, there is very little chance that we will reach them. This system will teach you how to overcome those doubts and fears and to live a life of abundance that is rich, fulfilling, and satisfying. Why would you want to continue living with a mindset of scarcity when a better life awaits you?

You, and only you, have the power to make these changes. As you will learn, it is simple but not always easy to make significant changes in your life. You must have a great deal of perseverance, determination, and passion in order to reach your goals. Yet these are all qualities that you already possess, but might have never believed yourself to have.

Anyone can change his or her life using this system. It works because it teaches you how to have complete control over all aspects of your life and, with that control, how to make your life better. You are about to discover your powerful new identity. The changes you make in your life using this system will affect others in your life as well. They, too, will benefit from the knowledge that you will possess after reading this book.

We all want something more out of life. Why not take the opportunity now to create a life you truly want. It is entirely up to you to change your habits and mindset so that you can begin to prosper. You will find yourself shedding all of the negativity that has built up in your life and simply throwing it away. This system can work wonders if you follow it; you could very quickly become someone that is truly free from his or her fears and doubts.

The use of statistics will help you to reach your goals. Statistical data, for our purposes here, is merely to be used as a guidepost for our journey. We can avoid becoming part of negative statistics by allowing the data to tell us what we should do. In this way, we can circumvent any potential problems by being prepared for them. We can predict what our conscious choices should be and act upon them without reservation.

Let me take you now to a place where you are the most powerful person in your life and the master of your destiny. Using the power of statistics to help us along, I welcome you to a world of choices and dreams. You can do it. This system will teach you how.

CHAPTER ONE: The Power Of Statistics

Habit is [our] sole comfort. We dislike doing without even unpleasant things to which we have become accustomed.

—Goethe

In many ways, we are slaves to our habits. We wake up each morning to our routines. We drink our first cup of coffee, walk the dog, and head to work. Throughout the day, these habits, these rituals, are played out in cities and towns throughout the world. Why do we have these habits and why do we allow them to affect us to such a degree? How did we learn to do the things that we take for granted every day? What is it that compels us to go to certain places for lunch or to jog on the same route each morning? These small habits can add up to quite a chunk of our lives if we really think about how much they affect us.

These habits are what we consider part of who we are, and they influence the neighborhood that we live in and the community that we represent. Habits are things that we do regularly without hardly any thought. We buy the same toothpaste each time we go to the store; we have a similar routine every day. These habits comfort us; they let us know that we are at home in our environments and that we somehow belong to our lives.

PARADIGMS

The sum of these habits is often referred to as a "paradigm." There are many different ways to describe a paradigm. There are paradigms of nature, paradigms of science, and even paradigms that encompass how a person thinks and believes. Each paradigm represents what we would call "inside the box" thinking. A paradigm symbolizes the habits and everyday mindset of individuals that have been learned or developed over time.

Following a certain paradigm could have a positive or negative effect upon your life. If you follow one that does not work for you, or has negative effects upon your life, then you can choose to change it. If the paradigm in question has positive effects, then you may choose to keep it. However, when we remain in the same paradigm, we risk losing our ability to make rational and free choices. This becomes our comfort zone and could be damaging to our health, finances, personal relationships, and to our potential for living more productive lives.

Following our established set of paradigms makes us feel safe and secure. When we do things in an old and familiar way, we like the good feelings and sense of well-being that accompanies these actions. We have trouble diverging from these paradigms because we, as human beings, resist any change that brings on feelings of insecurity. Even if our paradigms lead to negative results, we remain faithful to them, too afraid to abandon the ritual of our daily lives. Many people, for example, remain in a job that they hate because they are afraid of change. Also, there are individuals who remain in abusive relationships because they, too, are afraid; perhaps that is all they know.

It seems natural and easy to follow our current paradigms and rarely give any thought to the idea of change. Why should we trade safety and security for a new and frightening way of thinking? Why should we want to make choices for ourselves instead of going along with the status quo? Why shouldn't we simply follow our own paradigm, or "program," and wait for it to lead us to success?

Our current paradigms are comfortable, like old and soft blankets. We cling to them like children because many of us fear new or unknown things. We do not want to exchange comfort and stability for a new paradigm. That would mean adjusting to a new way of life and a new way of thinking, which can be daunting, especially if we view our lives as "not that bad." It is easy to sit back and allow our habits to control us rather than take responsibility for what happens in our lives. Changing our thoughts are as, or more, difficult than breaking any habit. St Augustine once said, "Habit, if not resisted, soon becomes necessity." Thus have our paradigms become necessities for us.

If we continue to follow paradigms that no longer work for us, we will soon believe that we are trapped and unable to make choices. This kind of thinking can seriously affect a person's life, stealing the joy of the present and hope for a better future. When we give up our free will to make choices, we give away the control we have over our lives to the circumstances and people around us. For example, many people will blame others for the situations they find themselves in. They will accuse their employers, their government, and anyone else that they can think of for their circumstances. Yet this is simply evidence that these people feel that they have no control. We allow our paradigms to control us, instead of creating new systems of belief for ourselves. People who think this way could be called victims. They blame other people or circumstances for their situation. About 95 per cent of people belong to this group; the remaining 5 per cent are people who realize that they themselves, through their choices, are responsible for their results in life. And if they do not like the results, then they are responsible for the changes necessary to get new and better results.

Clearly, following old, defunct ways of thinking and allowing them to control us is unproductive and destructive. Some people choose to follow a given path rather than to make conscious choices in their life. Yet, when a person does this, he or she risks the negative effects that this type of action might bring. It is far better to make choices that will affect your life than to allow a paradigm or mindset to control your actions.

One way that you can control your actions and make good decisions is through the use of statistics. Statistics is the understanding that your choices lead to certain outcomes that could be negative or positive; it is an understanding that tell you that you have a choice. If you understand how certain principles can be applied to create a better outcome, then you will have a better chance of getting a positive return on your investment – better than you will by continuing on your current path.

You can make decisions that will end in a "better return" for you if you first understand that you have the power to make these choices; they are not predetermined outcomes. You always have the opportunity to make different choices than the ones that currently fit into your paradigm. You are in control of your choices, and simply deciding on this assertion can be beneficial. Take, for example, the old story about the alligator and the bird.

THE ALLIGATOR AND THE BIRD

The story of the alligator and the bird has its origins in one of Aesop's fables, "The Scorpion and the Frog." The basic premise of the tale is that an alligator and a bird meet at the riverbank. The bird wishes to cross the river because, for some reason, it cannot fly. The alligator offers to give the bird a ride on his back. The bird protests, saying, "But, if I let you give me a ride across the river, who is to say that you will not eat me before we reach the other side?" The alligator manages to convince the bird that it will be safe to ride upon his back, and he promises that he will not eat him before they reach the other side of the river. The trusting bird decides to put his faith in the alligator and they set off across the river, the alligator with the bird upon his back. Midway across the river, the alligator snaps his jaws and begins to eat the bird. The bird protests again, saying, "How can you eat me? You promised that you would not and that I could trust you to take me across the river safely!" The alligator replies, "Why are you so surprised, little bird? You knew that I was an alligator when you hopped onto my back."

The moral of this tale is that we often find ourselves in positions that feel negative, unsafe, or downright unhappy because of the decisions we have made. Like that little bird, we have the opportunity to make choices that will benefit our lives. If we choose to trust the alligators of this world, or make unhealthy decisions, then we truly have no one to blame but ourselves. You, like the bird, always have the opportunity to make your own decisions, to shift your way of thinking, or to escape your old beliefs.

STATISTICS: CHOICES IN ACTION

One way that you can avoid remaining in a paradigm that is not working for you is to step back and take a look at your life from a statistical perspective. Statistics is a pure science, meaning that it takes a very straightforward and objective look at the facts. You encounter statistics every day. For example, you may have heard about studies noting the success rate of heart transplants or the average age of the people in your hometown; these reports are nothing more than statistics in action. They give you the possible outcome of a certain set of known variables. Note that I said possible. This means that it is a percentage probability, not a set outcome. You are not doomed to follow a certain path, and if you choose to take a different direction, then the variables change.

We know, for instance, that it is theoretically possible to run through a concrete wall without being harmed. If you tried enough times, eventually each atom in your body and in the concrete wall can be in the perfect position to allow you to run through it. However, we also know that this action is impractical, and its outcome improbable; it is just possible. Just because something is statistically possible does not mean that it makes good sense, such as attempting to run through a concrete wall! We must develop the ability to utilize the power of statistics with common sense in order to make changes in our lives that are positive and bring us what we desire.

THE POWER IS WITHIN YOU

The power to change your life lies within your own mind, and nothing prevents you from making changes in your life. Consequently, no one else can do this for you. You must find it within yourself to make choices that create a healthier, more positive lifestyle. This is when you can harness the power of statistics and make it work for you.

If you know, for example, that the divorce rate in the area where you live is at 50 per cent, you can take that information and, instead of feeling tethered to the statistical data, choose to change your attitudes and actions so you do not become trapped by them. Knowing the statistics allows you to make changes to avoid becoming part of that divorce percentage. You could choose to focus more on your family and on your relationship with your spouse. You could make the commitment to spend more time with your wife or husband, and to participate in more activities together. This will create an environment where you are more attuned to your spouse's wants and needs.

There are many ways that you could make your marriage better simply by knowing that 50 per cent of all marriages end in divorce. Knowledge is the first step because it makes you aware of the potential problems, thus allowing you to foresee and avoid them. It is only this constant state of awareness that will allow you to circumvent potential problems, and statistical data is often the only accurate source for the information.

A BETTER RETURN

Now that it is understood that you have the power to make choices and to break away from the old ways of thinking, it must also be understood that only the use of the power of statistics will grant you a "better return" in everything that you do. Armed with this knowledge, think of how far you could go and how much you could achieve. Making choices that are not governed by a particular paradigm can be both exciting and exhilarating. Just think: you have the power and knowledge to change your life now!

For example, when I was at the university, my statistics professor made all of the boys in the classroom stand along one wall and all the girls along the opposite wall. She told us to move forward, but only so far that the distance between us was divided by two. Then she asked us how many times we had to do this before the two groups met each other physically. The bright students, after some thinking, said that that that would never happen because half of a given distance would always would be greater than zero. This was correct. Therefore, no matter how many times you cut the distance in half, you will never reach your goal; the boys and girls would never meet at a common point. Our professor told us that the answer was right, but she also said, "After doing this seven times, you are close enough for all practical purposes."

This kind of combination of statistical knowledge and common sense is important. Remember the concrete wall. For all intents and purposes, a person cannot run through a concrete wall. In the same way, cutting the distance in half seven times in the above exercise will lead to two persons meeting at some point, for all practical purposes.

CHESS

The chessboard is the world; the pieces are the phenomena of the Universe; the rules of the game are what we call the laws of Nature, and the player on the other side is hidden from us.

—Thomas Huxley

Almost everyone remembers the story of Bobby Fischer. He was a young, eccentric chess player who disappeared from the chess scene in 1972. In what came to be known as the "Match of the Century," Bobby Fischer defeated Boris Spassky to become the eleventh world chess champion and a legend.

Imagine a rematch between Spassky and Fischer. This time, however, the stakes are different. The winner of the match must place one grain of rice on the first of the sixty-four black-and-white squares that make up the chess board. On the second square, he must place two grains of rice (double the amount of the first square). On the third, he must place four grains of rice; on the fourth, eight; and so on, until all of the squares are filled. It seems like a simple task, but if a person were to actually attempt this, he or she could never complete the task. Doubling the amount of rice each time would, in the end, amount to placing more rice on the chessboard than we can even comprehend! Theoretically, this can easily be understood. It is only when practicality comes into play that the task becomes impossible.

Statistics, therefore, can be of use to us theoretically, but it often falls short when it comes to common sense and the practicality of its use. Only the combination of this theoretical knowledge and the common-sense approach will lead to success.

For example, if there was a study published about crime rates on college campuses, the statistical data might reveal no crimes reported in the year 2007. However, is this really the case? Probably not. Just because a crime was never reported doesn't mean that it didn't happen. It simply means that it was never reported. This happens quite often. However, common sense tells us that all types of crimes actually *do* occur on college campuses. They just aren't reported. So, our common sense tells us that crimes occurred; even though they may not have been part of the statistical data, it is still possible that they occurred.

THE GOLDEN MEAN

Socrates is arguably the most well-known philosopher. He was the mentor of Plato and the person who gave us the Socratic method, in which one does not learn from simply memorizing or studying what is taught, but by experiencing knowledge through critical thinking and, thereby, gaining wisdom in the process. Socrates did not believe that one should write one's philosophical thoughts down. The task of commemorating Socrates' wisdom thus fell upon Plato.

In what is usually referred to as the Platonic dialogues, Socrates is the main character. He walks around the city of Athens, creating havoc wherever he goes. One of Socrates' favorite pastimes, you see, was to create situations in which people who believed themselves to be truly wise could be shown to be nothing more than fools.

In a lesser-known dialogue, *Laches,* the discussion centers on the virtue of courage. Socrates asks, "What is courage?" In the presence of many of the mightiest Athenian warriors, he might expect a fairly accurate answer. One of these men suggests that the virtue of courage might be defined as the ability to run onto the battlefield without any fear. Another suggests that, though he does not claim to know what courage might be, he understands cowardice as the opposite of courage. He then says that cowardice might be the presence of too much fear, the feeling one gets when he or she is much too afraid to run onto the battlefield. Finally, a third man suggests that true courage might lie somewhere between those two definitions. If courage, for the first warrior, is the ability to run onto the battlefield without any fear, and if cowardice, for the second soldier, is the inability to do the same, then couldn't the answer lie between those two extremes? In fact, wouldn't true courage mean that a person *is* afraid to run onto the battlefield but *does it anyway*?

It isn't really a very courageous act to run out into battle without any fear. Perhaps we could consider this pure insanity or a lack of emotion. This type of "courage" is not "true courage" because it does not require an inner struggle. If I am not afraid to walk across a room, does it require much courage for me to do so? The answer lies, instead, between the two extremes. It lies somewhere between a deficiency in courage (cowardice) and an excess of courage (no fear).

This formulation is called "the Golden Mean," and it was used across Athens during the time of Socrates, Plato, and Aristotle. It suggests that virtue, or that which is good, always lies between the extremes of excess and deficiency. The power of statistics, too, follows a similar principle. The virtue of the use of statistics in your life in order to get a "better return" is a balancing act between statistical facts and figures and a practical and common-sense approach. The ability to make choices, such as escaping or discarding a way of living that is not working for us, is the power we possess. This power can be harnessed through the use of statistics. We can avoid many pitfalls if we are able to see them instead of floundering around in the dark.

FREAKONOMICS

If you can question something that people really care about and find an answer that may surprise them – that is, if you can overturn the conventional wisdom – then you may have some luck

—Levitt and Dubner

Let us look at another example. In their book, *Freakonomics*, Steven D. Levitt and Stephen J. Dubner explore the use of statistics in the world of economics. Yet theirs is also a common-sense approach to looking at the world around us.

Beginning only with raw data and a question, these self-titled rogue economists set forth to point out normally overlooked conclusions that could be drawn from statistical data. The topics range from "What Do Schoolteachers and Sumo Wrestlers Have in Common?" to "Why Do Drug Dealers Still Live with Their Moms?" Each question is answered in a creative and unexpected way by using simple data.

In 2002, the No Child Left Behind Act was enacted by President George W. Bush. It was meant to ensure that each child in the United States learned everything he or she was meant to learn during their school years. Public school systems have been using standardized tests for years to determine whether or not children were getting a proper education. However, now these tests became even more important. Children were no longer allowed to be promoted to the next grade simply because of their age (social promotion). Students would now have to earn certain test scores in order to be promoted to the next grade. However, the news wasn't all good. Students would no longer be "left behind," but schools with low scores could now be penalized for students' low scores. Teachers were now responsible not only for their day-to-day duties, but also for high test scores. At stake were promotions, censure, and loss of large bonus payments.

This led to "teaching to the test," a practice in which teachers exclude all information from classroom instruction that might not be on the annual standardized tests. Students are no longer being taught what is important for them to learn, but are only being taught information that will show up on standardized tests. This also led to cheating.

Students weren't the ones cheating; this time, it was also the teachers. Teachers, worried about their future in the educational system, began to change blocks of incorrect answers on their students' test papers to correct answers. They had every incentive to do so – a secure future and additional pay. They were, of course, caught and promptly dismissed.

Shocking? Yes. However, could these rogue economists be right? Could there be connections between social and political decisions and statistics? Absolutely. If we use statistical data to inform our decisions, to assist us in either making the right decisions or avoiding harmful ones, we just might get a "better return." If we knew, for instance, how to stop terrorism or how to create a lasting love using statistics, shouldn't we utilize it to the best of our abilities?

For instance, let's look at the obesity epidemic among children in the United States. Obesity in children aged six to eleven has more than doubled in the last twenty years. It was only 7 per cent in 1980, but by 2004, it had grown to 18.8 per cent. Using this statistical data, we can stop this problem. We can make sure that our children eat healthy foods, drink clean water, and avoid too many snack foods that are bad for them. Another way to alleviate this growing epidemic is to encourage children to play outside more often instead of sitting in front of the television or playing video games.

We all have the power to make positive choices in our lives. If statistical data suggests that the mortality rate in a particular area is high, does that mean that it is a more dangerous place to live in? It could, on the contrary, mean that the population there is made up of older people. Does the statistical data also mean that, if we live in an area with high mortality rate, we are doomed to die at an earlier age?

Clearly, this is not the case. We can make choices to change our lives instead of resigning ourselves to becoming part of the statistical data. If we want to avoid death at an early age, we can take steps to make sure that we do not die early. We can eat healthy foods, take walks in the evening, control the types of chemicals we allow into our bodies, and keep healthy in many other ways. We can make choices that will change the outcome of our lives.

We can live longer lives than statistics might suggest by making healthy choices. We can have better personal relationships by using statistics to tell us what we should do. We can live better, more fulfilled and healthier lives by making good choices. It is only then that we will get a "better return."

CHAPTER TWO: How Statistics Can Be Misleading

You can't be distracted by the noise of misinformation.

—James Daly

Although we have already learned that statistics combined with common sense can be useful and help us to make good choices, it is important to note that some statistical data are not reliable. Sometimes, statistics can be misleading and thus distract from an otherwise positive path. Statistical data can be collected in an unethical or improper way, making the results of such research both incorrect and deceptive.

Although this chapter takes the issue of statistics seriously, it is important to learn what statistics can teach us. We must know how statistical data can be misused in order to gain a better picture of how to use it to our advantage. Even data that is misleading can be useful to us in this endeavor.

PARLIAMENT

In Norway, in the early 1930s, research about voter statistics was done before that year's election to Parliament. The results of this study showed that the conservative party would most likely win about 80 per cent of the representative seats. Everybody in Norway believed that there was something very wrong with these statistics, but what could that be?

Upon closer inspection, it turns out that the research had been done by randomly picking names from a phone book and asking these people how they would vote in the upcoming election. This might seem like a fair way to get a good idea of the potential results, but what it didn't take into account was the fact that most people did not own telephones during this time period in Norway. People who did own phones were much more likely to be members of the conservative party; therefore, the research was shown to be meaningless. Clearly, the results would have turned out very differently if the analysts had used some other method of research or been aware of the percentage of people in Norway who owned phones and were conservatives.

RADIO DIARIES

It is easy to see how results of studies and research can be affected by many variables. This same situation occurs in modern-day research on radio listenership. We all hear about how one particular radio station is more popular than the others. When a radio station is considered to be popular, it has a greater chance of selling ads to clients and generating more revenue. People are always more willing to invest their hard-earned monies on a radio station that they believe more people will listen to. However, many people argue that the means by which the ratings are determined is unfair and outdated.

When assessing radio station ratings, third-party companies will send out "diaries" to listeners. Each person, or household, keeps track of what radio station he or she listens to, for how long, and so forth. These diaries are then used to determine which radio stations are the most "popular." If the majority of the people listen to 109.2 "the Station," then "the Station" will be given the highest ratings for that period. This seems like a fair practice, right? The statistical data that is used to support the conclusion comes from actual listeners, and that should be the source for real numbers. But is it?

The fact is that the diaries are only given out in one or two parts of the city in question each time the ratings are determined. For example, if you live in the northern portion of a larger city, diaries would not be offered to your household each time a ratings period came up. The issue here is that different areas of town are distinct communities, and people who are like-minded often stick together out of comfort.

In other words, if you are a college student, you are most likely going to live in an area that mainly consists of other college students. College students are normally on a tight budget, so they will not live in the most expensive or affluent parts of town. Also, familiarity makes us feel good. Students attending a university or college are more likely to share a world view with others that are close to their age than, for example, people living in a retirement community. Most importantly, for our current purposes, people living in smaller communities within cities tend to share (on average) similar tastes, perhaps even in music.

So, what happens when the area selected for the distribution of the diaries also turns out to be a community of people with similar tastes, likes, and dislikes? What happens if no one who is given a diary likes country music, or urban music, or soft rock? You can see how easily the results could get skewed. If a particular community has the diaries, but does not, majority-wise, enjoy listening to country music, then the country music station will likely have lower ratings than it would during a rating period in which the diaries are given to a community where country music is much loved.

Clearly, a person could never trust this type of ratings system. It would be better if ratings were dependent on somehow monitoring the actual time individuals, not entire communities, spent listening to each station. Those who utilize this type of research believe that by purposefully giving the diaries to different communities within a city, that the results of such research would be fair and accurate.

JUST A RED HERRING

Misleading information can be compared to the idea of a "red herring." In the study of logic, a red herring is a defect in an argument in which the arguer attempts to lead the reader or listener off-course by introducing a completely irrelevant element. The reader or listener is drawn away from the original argument and follows, instead, the red herring.

The history of this phrase goes back to the practice of teaching British hunting dogs to follow specific scents. Trainers would attempt to throw the dogs off-track by using the strongest scent possible – that of the red herring (one of the strongest smelling fish available). Once a red herring was dragged across the ground, only the best dogs would follow the original scent.

Perhaps you know the concept of a red herring best from the stories of Sir Arthur Conan Doyle about one Mr Sherlock Holmes. In the Sherlock Holmes stories, a character, event, or misplaced "clue" could serve to draw the famous detective's cohort, Dr Watson, off the track of the real criminal. Most mystery stories include red herrings; they are plot devices that draw the audience off-track so that the ending is a surprise.

For example, let's say that someone wants to argue for the death penalty in crimes of murder. That person may argue that the death penalty is a bad thing because the victim's family (who are also tax payers) would have to pay for the murderer to be housed, fed, and taken care of. However, this argument includes a red herring and is, therefore, a bad argument. The fact that a murderer might have to be housed and fed by the prison system has very little to do with any argument about the death penalty. This is simply a slightly related topic, meant to draw the reader's attention away from the real argument: whether or not the death penalty should be used in cases of murder. Often times a red herring will be a far more exciting subject than the original topic, and, therefore, it can easily draw the participants off-track.

HEART TROUBLE

Often, statistical data can be as misleading as a red herring. If we do not understand how to weed out the good data from the bad, we can easily become confused and follow our own red herrings. For example, there was a time when people believed that the symptoms for a heart attack were the same for everyone – for people of different ethnicities, for men and women, and for persons of every age. We all remember that the symptoms to look for in a heart attack are things such as radiating pain in the left arm, chest pains, and so forth.

However, the general public did not know that all of the research on heart attack symptoms had been done on Caucasian males over the age of fifty. How could this research apply to everyone? It could not. Men and women experience very different heart attack symptoms, as do people of different ethnicities. The well-known statistical data on heart attack symptoms was not trustworthy because the research only focused on white males over fifty. The data here could not possibly be useful to us. In this case, we cannot even look to statistics to determine how we can change our future.

TUSKEGEE, ALABAMA

Research doesn't always have to be looked at with a critical eye simply because the process might have been unfair. The methods of collecting statistical data can also be unethical. This was certainly the issue in the case of the Tuskegee Syphilis Study.

This study is a prime example of an unethical research project, ruining any possible knowledge that could have been gained for use in the future. Three hundred and ninety-nine African American male sharecroppers who had already contracted syphilis were studied by the Public Health Service of the United States of America. The scientists working in Tuskegee, Alabama, with these men claimed that the study was meant to track the course of the disease, even though that information was already well known.

From 1932 to 1972, the participants in this study were observed. They were never given medication or much treatment. The Public Health Service did give them a place to stay and a small amount of money for their participation, but, on the whole, the sharecroppers were simply allowed to die. Even when penicillin became a common and safe treatment for syphilis, these men were not given access to it. They were told that they had "bad blood," not syphilis.

For 40 years, the Public Health Service of the United States of America lied to nearly 400 people. Certainly, the results of this study should not be trusted. Any research or statistical data discovered under disturbing circumstances can only lead to misinformation.

THE NUREMBERG CODE

During World War II, Nazi scientists experimented on many human beings, treating them like laboratory mice. The experiments that occurred in places such as Auschwitz are too horrible to describe. It is of great importance, however, that we note that the human subjects were not willing participants – they were prisoners and were forced to participate.

Out of the war crimes of the Nazis came what is known as the Nuremberg Code. This code states that any research done on human subjects will have very strict rules and regulations to protect the subject. Never again would the kind of unethical biomedical research that Nazi scientists participated in be allowed to occur. The American Medical Association even adopted the same spirit of the Nuremberg Code in its policy on research involving human beings.

There is, in many circles, an ongoing debate as to whether or not to use the "fruits" of Nazi biomedical research in the present. One side argues that no one can prevent the suffering of these human subjects as it has already happened. Therefore, we should use the statistical data gained from these experiments, because to do otherwise would be disrespectful to those who died as unwilling human subjects. The other side of the discussion argues that this research should never be used, that it should remain unknown, and that it should stand as a testament to those who gave their lives. Which is the correct perspective?

For example, one particular set of experiment was designed to determine how to deal with the ravages of hypothermia. Nazi scientists placed their prisoners in freezing water until they became hypothermic. They then attempted, through various ways, to revive those people and to return their body temperature to normal. You see, the Nazis were worried about their airmen; hypothermia was a serious problem for soldiers during that time. They discovered many interesting things about the treatment of hypothermia. And despite their lack of ethics, they kept immaculate records of all the experiments they performed.

The data is certainly reliable in the sense that it was collected with no regard for the human beings involved. However, does the very fact that it was unethical mean that it cannot be used to help others who have hypothermia today? If the Nazis had found a cure for cancer through horrible experiments conducted on human beings, could we deny current cancer patients the cure simply because of the ethics of the researcher?

ON THE LIGHTER SIDE

USA Today has come out with a new survey: apparently, three out of every four people make up 75 per cent of the population.

—David Letterman

Marilyn Vos Savant ("Ask Marilyn" of *Parade Magazine*) agrees that statistics in many cases can be misleading. She points out a few interesting and entertaining sets of statistics. Evidently, for instance, seven per cent of Americans think "espresso" is an "overnight delivery system." As Vos Savant maintains, this is most likely due to the type of survey questions asked in order to attain statistical data. In fact, she discovered that the questions in the survey about espresso were multiple-choice questions, with "overnight delivery service" as an option. She is also quick to note that a growing number of people in the United States are not native English speakers. For some, "espresso" might sound like Italian for "quick" or Spanish for "special delivery."

Another statistic that Vos Savant talks about is the fact that people in western Germany claiming to have indigestion has gone up by five per cent since 1989 (when the Berlin wall fell). What could be the meaning of this data? Does this mean that western Germans are more nervous or upset than they used to be before the Berlin wall was brought down? What is the correlation or relationship between these results and the information they are meant to convey?

Certainly there are several reasons why seven per cent of Americans believe that espresso refers to a delivery service or why western Germans might have a higher rate of indigestion than they had in 1989. However, this official data does not really show that one thing caused another – there is no evidence of causation, only correlation. The two are related, but one thing did not necessarily lead to the other. This is an important distinction to make when analyzing statistics.

Are there really seven per cent of Americans who believe that espresso means something different than it does? Yes. Yet did the types of questions asked skew the results? We would have to say yes to that question as well. So, are the statistics reliable? Well, maybe it's reliable for a laugh, but not much more than that.

And what about the Germans with indigestion? Did the fall of the Berlin wall in 1989 cause a five per cent increase in indigestion in western Germans? The question we must ask ourselves is whether or not the fall of the Berlin wall *caused* the indigestion of more western Germans. There probably isn't a cause-and-effect relationship there either. Still, these statistics could easily convince someone that the destruction of certain structures can cause indigestion.

COMMON SENSE

Henry Clay (statesman and congressman of the United States) once said that "statistics are no substitute for judgment." It is difficult to navigate the treacherous waters of statistics. Here, using common sense is of the utmost importance. We have to think critically about the statistics we depend upon to help us to make good decisions.

It is not enough to read statistical data; it is necessary to analyze it and to understand that not all statistical data is reliable. Some studies, such as the Parliament and radio diaries, are simply mistakes and not intentionally misleading. Other studies, such as the Tuskegee Syphilis Study and the biomedical research of Nazi scientists, are unethical. A few statistics are even funny and can make us smile, but we don't take them seriously.

Nearly every day, we are pounded with statistics. In the early morning news, we hear statistics on crime, on population growth, on global warming, and on how many years in prison celebrities will spend. If you can learn how to distinguish useful statistics from outright deception, you can then steer clear of the problems that go along with misleading statistics and keep yourself from allowing it to lead you in the wrong direction.

Misleading statistics can, however, be helpful to us in the pursuit of discovering their power and influence in our lives. If you are aware that some statistics are deceptive, then you will be better prepared to determine which are truly leading you in the wrong direction. Determining which statistics to trust is part of arriving at decisions to get a better result. Distinguishing these nuances is part of the larger picture – your decisions will be enhanced by this knowledge.

CHAPTER THREE: The Best Return

People of mediocre ability sometimes achieve outstanding success because they don't know when to quit. Most people succeed because they are determined to.

—George E. Allen

RESOLUTIONS

It is important to know that many statistics can be misleading; however, it is also imperative that you know how to make choices using statistical data as your guide. Let us remember the example about divorce rates. These are not simple statistics; they are bits of information that can allow a person to make a choice about their lives. We all want to have better relationships, make more time for our families, and do the things we enjoy. Simply being aware of the data and using it to your advantage can help you to attain whichever goals you set out to reach.

Making a list of goals at the beginning of any endeavor is helpful. Some of us make lists of goals at the beginning of each year. Others set goals when striving for a particular achievement. Either way, writing down your goals solidifies them in your mind and helps you to stick to them. For example, you might want to save money, make your relationships last longer or be more satisfying, or simply improve your quality of life. Depending on what is important to you, you can find the appropriate statistical information to help you reach these goals in an effective and efficient manner.

THE "SUPERSIZE" FACTOR

A Pizza Hut! A Pizza Hut! Kentucky Fried Chicken and a Pizza Hut! A Pizza Hut! A Pizza Hut! Kentucky Fried Chicken and a Pizza Hut! McDonald's! McDonald's! Kentucky Fried Chicken and a Pizza Hut! McDonald's! McDonald's! Kentucky Fried Chicken and a Pizza Hut! I

like food! I like food! Kentucky Fried Chicken and a Pizza Hut! You like food! You like food! Kentucky Fried Chicken and a Pizza Hut!

—**Actual Children's Song**

Morgan Spurlock, a documentary filmmaker, decided to prove whether there was a role that personal responsibility plays in our decision making. He read a news story involving two girls who sued McDonald's for making them obese. Most of us know that obesity is a growing epidemic in the United States and around the world, especially in westernized cultures. Spurlock wanted to know whether fast food, specifically that served at McDonald's, was to blame for the epidemic. Was McDonald's food really that dangerous, or should personal responsibility be taken into account (that of the girls or their parents)?

Spurlock was determined to find the answers, much to the chagrin of his girlfriend, a vegan chef. He decided to conduct an experiment in which he would eat only McDonald's food for thirty days in a row. Monitoring his health would be important, so he enlisted the help of several doctors and nutritionists to keep track of the possible changes in his body during that month.

Of course, there were rules to follow. Spurlock was not allowed to eat any product that was not sold at a McDonald's franchise, including bottled water. For thirty days, he was to buy all of his meals at McDonald's and to eat there three times per day. If the cashier asked him if he wanted to supersize his meal, he had to say yes (hence the title of the documentary: *Supersize Me*). In addition to all of this, he also had to eat every item listed on the McDonald's menu at least once before moving on to the next thing.

Being the boyfriend, and now husband, of a vegan chef, Spurlock was in fantastic health before this experiment began. He ate a variety of foods during the day, but always ate a vegan meal in the evenings. Spurlock began the project weighing in at about 185 pounds and standing 6 feet, 2 inches tall. He ended the experiment at 210.

From the first meal that Spurlock eats at McDonald's (supersized, of course), we realize that he has gotten himself into far more trouble than we first thought. He says, while consuming a double quarter-pounder supersize meal, "See, now's the time of the meal when you start getting the McStomachache. You start getting the McTummy. You get the McGurgles in there. You get the McBrick, then you get the McStomachache. Right now I've got some McGas that's rockin'. My arms... I feel like I've got some McSweats goin'. My arms got the McTwitches going in here from all the sugar that's going in my body right now. I'm feeling a little McCrazy." He feels terrible from the very outset. In the end, he gains 24.5 lbs (11.1 kg) and his Body Mass Index increases from 23.2 (healthy range) to 27 (considered overweight). All in all, the experiment proved that perhaps eating a steady diet of any fast food products was not good for health.

Spurlock's team of physicians got worried when his liver enzymes rose dramatically and he had heart palpitations. They tried to convince him to abandon the project, so worried were they about his health. He declined and forged on, but it took him months after the project to lose all the weight he gained and to return to a healthy state.

CORPORATION GAMES

The following question then arises: what does Spurlock's endeavor prove? Corporate executives at McDonald's have always maintained that their food is not meant to be eaten on a regular basis, three times a day. They also argue that people ought to know that fast food is not very good for them, that they ought to take personal responsibility for their eating habits and stop blaming corporate America.

There are people in various countries who have tried to duplicate Spurlock's project, with different results. These people, however, did not perform the experiment to the same degree as Spurlock. Some of them claimed to *lose* weight on a fast food diet. However, what it comes down to is making the right choices for our lives.

This is where personal responsibility for the choices we make comes into play. If we know, or at least have a good idea, that fast food is not the best food to eat on a regular basis, then why do so? We can simply choose not to eat fast food often enough for it to negatively affect our health.

Furthermore, if we have access to data that proves that fast food can be harmful if consumed in large and irresponsible quantities (Morgan Spurlock's film for one), then we can use that data and make the choice not to eat it! It really is that simple. If we want the best return, all we have to do is *decide* to participate or not in certain things that we know are bad for us.

I know that it's not easy. This is why *Supersize Me* hits home. Many of us are just too busy for preparing home-cooked meals. We have jobs, children, and so forth that get in the way. And if everyone responsible for preparing the evening meal is tired at the end of the day, a trip to the local fast food joint doesn't seem unreasonable. Very few people truly enjoy cooking, but we all have to eat. Fast food is convenient, relatively inexpensive, and, let's not forget the most important part, tastes pretty good.

However, we know that obesity, especially in children, has become a very important health issue. Some researchers have even argued that obesity is more deadly than alcohol abuse or smoking. It can cause a range of problems, even at an early age. For instance, obesity can lead to heart problems, liver issues, pseudo tumors in the brain, diabetes, and a litany of other disorders and diseases. Why take the risk?

Again, most of us take the risk for the sake of convenience. Yet, armed with the data, you have the power to stop thinking so much about convenience and start thinking about what's right for you and for your family. This is, we all know, easier said than done. The important thing is to realize that this type of habit is just that – another paradigm to break free from. You can use statistics on the dangers of fast food to assist you in making a paradigm shift, which will only occur through hard work and dedication.

A LITTLE GUIDANCE

Statistical data is merely a guide. It is the choices we make, based on the data and a good dose of common sense, that are really the most important. I once read that Americans consume more beef than any other country in the world. People in the United States eat so much beef that if they were to decrease their meat consumption by 10 per cent, the grain used to feed the cattle that are eventually slaughtered could feed more than 60 billion people. That's quite a lot of grain. In fact, with those statistics in mind, a person could consider cows to be protein factories working in reverse. It takes more grain to feed the cattle Americans eat than the amount of protein our bodies get from eating the beef itself!

MALTHUSIAN ECONOMICS

Thomas Malthus would disagree. He was a successful economist that fell out of favor with his peers (and with the general public) because of what people now refer to as "Malthusian Economics." Long before it was a reality, Malthus predicted that world hunger would get worse. He argued that the problem was very simple; in nature, animal populations whose food source decreases die off. When the food supply returns to a normal state, the population grows accordingly.

Malthus stated that the way to avoid world hunger was never to feed a starving population. He wanted the populations of the world that had excess food to stop giving it away to people that were starving. You can see now why he was so unpopular. Human beings are not the same as rabbits and foxes. We like to believe we are on the top rung of the food chain.

Yet Malthus had an unexpectedly humanitarian aim in pointing out these unsavory truths. He knew that if a starving population was fed, they would continue to grow until their population that was once manageable (other countries or people can afford to give them their surplus without much trouble) doubled and then tripled. This would create an entirely new population of starving people.

Whether you agree with Malthus or not, you can clearly see his point. Why cause trouble for future generations of the starving population? Wouldn't it be better that fewer people perished of starvation? However, it is very difficult to escape our paradigms on this subject and to imagine that any human being could ever purposefully allow another to starve to death.

Malthusian Economics is one subject in which the statistical data is clear: feed starving people and you create more starving people who will suffer. Leave nature to its own devices; allow the population to ebb and flow with the availability of food and their ability to produce it. Nevertheless, knowing the statistics about starving populations does not make it any easier to make a decision. Who could decide that some people die while help is just around the corner? Malthus would maintain, however, that making the choice not to save a starving population would save the most lives in the long run. For most of us, this choice would be difficult to make, to say the least. Clearly, statistical data can guide us, but it does not always afford us the easiest decisions.

ADHD (ATTENTION DEFICIT / HYPERACTIVITY DISORDER): WHAT WE CAN LEARN

The American Psychiatric Association estimates that somewhere between 3–5 per cent of all school-aged children have been diagnosed with ADHD. It has also been said that ADHD is one of the most commonly misdiagnosed disorders in modern-day medical practice. Often, children who do not suffer from ADHD, but are simply a bit rowdy or energetic, are misdiagnosed with this condition and immediately placed on various drugs.

For a child to be correctly diagnosed with ADHD, several factors come into play. The child must be hyper and exhibit an inability to pay attention both at school and at home. Oftentimes, students who are actually gifted are misdiagnosed with this condition because they are bored with subject matter that they already understand, but which their teacher must review for those who do not understand it yet. Part of the key to determining whether or not your child has ADHD is comparing their behavior at home and at school. If the "ADHD behavior" is evident in both environments, then it is more likely that there is an issue with ADHD.

Children with ADHD are often difficult to control, and they exhibit behaviors that can be misunderstood. The medication that they are forced to take can make their daily routine a tremendous struggle, including their eating and sleeping habits. A child with ADHD is sometimes mistaken for one with severe discipline problems, instead of a child with a serious condition. It can be very difficult for families with ADHD children to live normal lives due to the medication and schedule that they must live under. Parents of children with ADHD find themselves constantly defending their children's actions, and this situation can stigmatize the children.

Many years ago, ADHD did not even exist. Hyperactive children were disciplined, given more challenging work, or handled in some other way. However, because of overdiagnosis, it is now the case that children who actually suffer from ADHD are lost in the crowd of others who may not actually have the disorder.

For instance, a child who actually suffers from ADHD will respond differently to medication than a child who has been misdiagnosed with the disorder. Teachers cease to understand why one child cannot "control" his or her disorder while another can. This leads to complaints of discipline problems and an array of other issues that could affect the child who has been correctly diagnosed with ADHD. Parents are forced to continually explain why their child must be treated as one with a medical condition, not a behavior disorder.

writing being only a few of the subjects taught. The Ezell and Levy article points out the importance of arts programs in juvenile correction facilities, not only while the offenders are imprisoned, but also after they return to the real world.

Statistics say that these teens won't make it in the real world after they leave prison. Most people believe that they will not be able to function and end up living a life of crime. Yet, knowing what these statistics say, we can do something about that. Just as Susan Warner did, we can fund and begin arts programs for these teenagers and help them to grow and to find their place in the world. The arts is a way for them to connect with other, positive influences than the ones they grew up or hung around with, and for something wonderful to happen in the process: learn self-confidence and be better able to maneuver in their world.

Clearly, the use of a practical approach to statistics can assist those in juvenile detention centers escape from their unhealthy paradigm. It takes a lot of work and dedication, but it can be done. It is not easy to get away from a harmful paradigm that is comfortable but bad for us. However, these youngsters had a choice just like you and I do. They chose to be part of something larger, something that is good for them and will help them when they get out into the real world. They chose to make a paradigm shift. And those that helped them chose to get involved in such a way that they could make that happen.

Again, it's not a simple task to make a paradigm shift. We feel safe, and continuing to do what we have been doing becomes habit, becomes easy to do. Yet, like the people involved in the arts program, we know that it can be done. Those young adults were able to do it; they were able to make choices that were not in-line with the statistics. If statistics told these kids that they were doomed to a life of crime and misery, then these teens said the statistics, "No way!" They used the statistics in order to make a change. They *decided* that they didn't want to end up that way, and most of them didn't.

What can we learn from this? The case of ADHD is a prime ex correct statistics (3–5 per cent of all children really are diagn ADHD) that is of little use to parents who suspect that their ch from this disorder. Yet we can use these statistics to protect ou from either being misdiagnosed or treated unfairly in the class a correct diagnosis has been made. The more informed our c better they become. The more we know, the clearer the choice i

THE BEST RETURN: AN UNEXPEC COMBINATION

Very few of us think about the juvenile correction system general idea that it is a different justice system than the on adults, but we know little about it. As it turns out, the tradit justice system has been changing. Instead of creating an based on rehabilitation, juvenile justice policies have begun on punishing juvenile offenders for their crimes. Is this sy No. The threat of punitive actions has not changed the juve So, what's to be done?

In an article entitled "An Evaluation of an Arts Program f Juvenile Offenders," Mark Ezell and Michelle Levy delve in ask what difference arts programs make to these children also wonder what kinds of surprises they might find alon

Ms Susan Warner had an idea about what to do. In 1 program called "A Changed World," in which the arts w juvenile offenders and to assist them in forging connecti world, increasing self-esteem, and making better gene their lives. The focus here was to move away from puni criminals and to help them instead.

Different types of artistic work were taught in the consisted of several short workshops – sculpture, pai

If children in juvenile detention centers can shift their paradigms and make their lives better, why can't we? It involves making a simple choice. If I know that smoking cigarettes will harm me, then I should choose not to smoke; if I know that working in a specific industry will be harmful to my health, I should try to get out of that industry.

A friend of mine – let's call her Julia – was once an important record producer living in an expensive area of the United States. She had a very busy life, a huge house, and much wealth. Yet Julia was not happy. She wanted something more out of life, but wasn't sure what that was. Julia thought about it for a while; what was making her unhappy? She wasn't sure that she even knew. She had been caught up in the paradigm of her life and career for so long that she could no longer see why it wasn't working or if it was even working.

Eventually, the answer came to her: she would simply *decide* to change her life. She moved to the desert; it was quiet and beautiful and she raised horses. Leaving her high-paying job wasn't easy. She had to give up a good deal of prestige and some of her money, but in the end it made her happy. She realized that all it takes is a simple decision; we can all decide to make our lives better by *simply deciding to do so.*

Julia also realized something that most of us forget: that we are responsible for changing our lives and our circumstances by simply deciding to do so. Sure, you have to work at it. Julia wasn't able to leave her career immediately after deciding to do so. She had to make both subtle and large changes in order to create the life she wanted. And it was a creative process, completely directed by Julia herself.

She determined what she wanted out of life (a more relaxed job, a place that had clean air and sunshine, more time for her children, and a less stressful life in general), and then she went out and made it happen. Carefully, she traded the old and hectic job for a more peaceful one, and the huge house on the east coast of the United States for a ranch. She made the decision to change her life, to initiate a paradigm shift, and then did it.

We can all learn quite a bit from Julia. Most of us are not very comfortable with completely changing our lives, even if we know it will be for the better. We go to work every morning in the same office, in the same building, and in the same situation every day. But if it makes us miserable, we should stop. This seems obvious, but few of us have the courage to stop.

Albert Einstein once defined insanity as doing the same thing over and over and expecting a different result. This is what we do every day! Our course of action seems simple when put this way: to get the best return, look at the information and use it to make choices to obtain not only a different result, but a *better* return.

CHAPTER FOUR: Beating the Odds

Failure will never overtake me if my determination to succeed is strong enough.

—Og Mandino

REFUSING TO LIVE WITH THE OUTCOME

You do not have to live with the outcome that statistical data projects; you can change it. This is what I will now refer to as "beating the odds." In all parts of your life, you have to make decisions every day. You start making these decisions when you wake up in the morning. The first two are: should I get out of bed now, or should I wait until later? Shall I take a shower or not? The day goes on in that vein. However, making a choice will give you better results than following the usual routine. This is the key to using statistics to make your life better and to help you to escape the paradigms that have kept you from reaching your goals.

Making choices can be difficult. You must be aware of many different factors in order to make the correct decision. All of us encounter choices that we must make every day. Not all of them are as simple as deciding whether or not we will get out of bed at a certain time. Some of them are very complicated.

Many of us may have to choose which school or health care provider is best for our children. We may have to choose whether or not we want to accept a job that pays less, but gives us more time to spend with our families and is, therefore, much more valuable to us. Difficult decisions constantly surround us; we wonder what is best for us to do. What will make our lives better and, more than that, what do we want out of life? What are our goals?

If we can make choices that will lead us to our goals, whatever they may be, then we can "beat the odds" by changing what statistical data predicts. Creating a set of goals is quite a task; we must decide not only what we want, but how we will get it in the end.

Very few of us take the time to write down our goals, but we keep them in the forefront of our minds. In fact, we know that although five per cent of people actually have goals, only three per cent bother to write them down. Yet this is not enough. Goals must be specific, measurable, and have a time limit. This is the true definition of a goal. For example, "I want to earn $50,000 a year from my bakery business as of April 5, 2008," is an example of a proper goal. Constant awareness of our goals will keep us working toward them and away from paradigms that are not working.

RESOLUTIONS

For example, a friend of mine frames her yearly goals and places them in a place where she sees them everyday. This reminds her of what she wants to accomplish; every morning she looks at these goals and is reminded of what she wants out of life. However, top-of-mind awareness is not enough. Each of us needs a guide; we all want to know how to reach our goals and how to move away from unhealthy habits and thought processes. This is another way in which looking at statistical data and changing, through conscious choices, the results for our own lives can work for us. Making choices that are informed by statistical data will give us the "best return."

PIECES OF PIE

This idea is applicable to both the small and large parts of your life. Think of life as a large, delicious pie. Each piece is wonderful and important, and it is an entity in itself. Yet each piece makes up the whole – the whole of life. Therefore, we can think of the larger part of life (the whole pie) as being made up of smaller portions, each as mouth-watering as the next. I remember what I was once told: forget the minutes; take care of the seconds in life because these are the parts that your life is built from.

In other words, if we take care of the small parts, the larger ones will take care of themselves. Also, perhaps we should take better care of the smaller portions, like choices that we make, because they are the building blocks of our lives and will help us to create new ways of thinking and living that are more in line with our eventual goals. Through conscious choices, you can "beat the odds" and change what happens in your life.

The smaller portions of life consist of decisions that we make every day and hardly think about. This continues throughout our day. We have to start making decisions from the moment we wake up in the morning. Should we take the time to make something healthy for breakfast, to make something easy, or to skip breakfast completely? These small decisions are equally important because they add up to the sum of our lives.

The smaller choices we make give us our results. If we set goals and want to keep them, it doesn't seem as overwhelming to make smaller choices as it may to make larger choices. Once your goals are set, you should work toward them, but you don't have to do it all at once. Trying to reach every goal that we set would be a daunting task. No one could reach all of their goals in one day. It takes time and hard work to remove yourself from a confining belief system. Setting goals is only one part of the process; working to make those goals a reality is the next step.

THE "LAW OF SMALL STEPS"

Little by little, one walks far.

—Peruvian proverb

A friend of mine, who is a general in the Army, always said that in trying to reach a goal – any goal – you must always take small steps. Even if you could take one large step and reach the goal, it is always better to take several small steps.

If we take care of reaching smaller goals and making smaller choices, then the larger picture will take care of itself. As the seconds take care of the minutes, so do the smaller choices take care of the larger ones. If we make a concerted effort to make choices, then we can reach our goals more easily.

GOALS: AN EDUCATIONAL EXAMPLE

For instance, if my eventual goal is to get the education that I need in order to get the job that I want, I will have to work toward it. I cannot simply say to myself, "I will get the education that I need." I have to take care of the smaller things too. Getting good grades before entering university, taking tests and doing well on them, and keeping track of all of the course work that must be completed are the smaller decisions that have to be made. If, at the end of the term, I have several tasks to complete, then I must work hard and apply myself to get them done and eventually reach my goal.

Making choices on a regular basis, both on small and large matters, is what it takes to reach goals. Continuing with the above example, if I decide that I am too tired to do the work necessary, then I will not reach my eventual goal of graduating. If I determine that it is better for me to have fun or to waste my time doing other things, then that is my choice. However, choosing to complete the necessary work is also a conscious choice, one that will lead me to my goal. Choosing to remain organized and attentive to my responsibilities will end in happiness and a sense of accomplishment that completing goals can bring; it will also end in me reaching the eventual goal of completing an education.

There were many individuals that I knew in school who did not do this. They were amazed by the level of organization and dedication I had while accomplishing my goals. However, it was very simple for me; do the work, do it well, and you will reach your goals. This is where personal responsibility comes into play. Using statistics, knowing what my goals are, and then making choices that will lead me to those goals is up to me. I must make a concerted effort to make good decisions in everything that I do so that I can get a better return.

"AN OUNCE OF PREVENTION..."

This is the case with almost everything we encounter in life, from cars to relationships: part of getting a better return is knowing that it is always easier and better to prevent something from breaking down than having to repair something after the fact. I truly believe the old saying that goes like this: an ounce of prevention is worth a pound of cure. Many people will also say that something that is worth doing is worth doing right the first time instead of wasting time trying to fix whatever it is that is wrong afterward.

When it comes to cars, for example, we know that there are several problems that we can prevent if we simply pay close attention to its upkeep. A car can last a long time if the regularly scheduled maintenance, such as oil change, is performed. Air filters must be checked and, if clogged, replaced. It is also a good idea to have your car's brakes inspected and the coolant level and tire pressure checked. Most of the automobile problems can be avoided with just a little bit of care and precaution.

When it comes to relationships, there are tools, such as being considerate and patient, that we can use to avoid arguments with significant others or spouses. Many problems within a marriage, for example, can be prevented by the partners simply listening to each other and compromising. Again, if we know what the potential problems are in our lives, we can avoid them through careful planning and deliberate choices that are tailored to our goals. If each decision we make is geared toward reaching the goals we have set, then our lives can only get better.

If we can prevent our children from eating unhealthy foods that will make them obese and expose them to all sorts of health issues, then we should. In fact, healthy eating habits learned in early childhood have proven to help children to have healthy eating habits when they reach adulthood.

In the case of home ownership, the same rules apply. There are many ways that we can prevent our homes from deteriorating, such as keeping the roof in good shape, making improvements, and generally maintaining the house and yard. If a house is kept in good condition, then it will be worth more money and remain a good investment for the future.

Let us consider a situation in which a person is having financial difficulty. Most people depend upon credit in order to live. With oil prices rising daily all around the globe, food prices have also gone up. Trucks, which use this type of fuel, are unable to deliver groceries to stores as efficiently, and food prices go up as well. What can a person do? Budgeting is one solution to several financial problems. If you can sit down and write out a budget for the month, considering both your income and expenses, then you can live more efficiently; you can avoid getting into debt or making your financial situation worse. It seems as if this is too simple a solution, but it really does work. You will always know how much money you have and how much you can spend on any given thing without having to worry about it.

I will never forget my first boss, Rolf. He owned a garage that dealt mostly with car tires. At that time I was just a small boy of eight living next to his garage. The garage was, for me, a very exciting place, and Rolf put me in action doing simple things such as sweeping the floor, cleaning, and running small errands to the shop for him. At the end of the week I would get my salary. At that time, in the late 1950s, I would get something like half a dollar. For a little boy of eight, that was a lot of money. After giving me the money, Rolf *always* asked, "How much are you going to save this week?" Later in my life it became a habit for me to save some money every time I got my salary. Because of this habit I have some reserve money for times when a sudden need might pop up. This gives me a feeling of security, and I can do things without asking for help.

In other words, as with all things in life, repairing the damage is much more difficult than preventing it in the first place. We can use statistics to our advantage and make our lives better through deliberate decisions that will lead us to completing our goals. In the next several chapters, we will take a detailed look at many such situations and see how we you can use the power of statistics to make our life better in our personal relationships, our finances, and our professional and everyday life.

In an office or corporate setting, it is difficult to convince people with varying personalities to get along. There are several types of difficult coworkers. There are those that always say yes to everything the boss says, even though they have no idea how to do the tasks that they are assigned. In addition, there are those coworkers who believe themselves to *be* the boss – they believe that everyone in the office works for them and is completely incompetent.

With a few simple tools, however, these situations can be controlled. Regular meetings and concrete job descriptions are only two suggestions for this particular problem. For the boss, however, the situation may require a few very deliberate choices to make the business run more smoothly; the boss might even find it necessary to speak to a lazy worker in private or to fire an employee who has been branded a "troublemaker."

You must always remember how important it is to make deliberate decisions. Every day, we must choose to act and make decisions in a way that will lead us to our specific goals. In fact, it is a struggle not unlike that of an addict.

ADDICTION: THE MOST DANGEROUS PARADIGM

Drug addiction is a prime example of being trapped in a paradigm that is not working. It is an issue for many people around the globe and is, in fact, still a growing problem. People who are close to addicts always wonder why the addicts do not simply stop destroying themselves with drugs. How could someone, they ask, live their lives this way, stuck in a particular way of thinking and being that is so devastating to themselves and everyone around them?

Yet a drug addict is nothing more than someone stuck in the paradigm of drug addiction. Addicts do not see a way to make choices about their lives that will make their situation better; all they see is the opportunity for their next high. Oftentimes this situation ends in the complete destruction of their relationships with family and friends. Many times it means an early death for the addict. Drug addiction affects every aspect of the user's life. People begin to lose faith and trust in the person addicted to drugs and give up on helping the addict entirely, knowing that it is sometimes too painful to help someone who refuses any assistance.

Breaking out of this manner of thinking is difficult. Addicts must make a choice to get help where they can; they must make a serious commitment to reach the goal of being drug-free. The road to recovery is not easy, and very few former addicts ever get their full lives back.

Clearly, according to the logic in this book, it would be easier never to have become addicted to drugs in the first place than to have to deal with overcoming an addiction. Still, it takes a lot of courage to get through something like being an addict. Perhaps we can take inspiration from people who have beaten their addictions.

As mentioned before, a person's addiction to drugs can be compared to a paradigm or a way of thinking that is not healthy or beneficial. In order to escape that box, to leave that confined belief system, it is necessary to do some serious work. We will have to make a serious effort. All of this can be quite frightening, but if we use knowledge, specifically the knowledge of statistics, to either avoid problems or overcome them, we will be successful.

In fact, we are much like addicts in that we are addicted to our comfortable ways of life. We cling to old belief systems, hardly ever thinking that we can escape a miserable situation. But these systems do not work for us; we want a better return, not just an "okay" one. Everyone wants success in some, if not all, areas of his or her life.

Like the recovering addict, we must make choices that will lead us to our goals. For the addict, the choice might be to enter into a treatment program. For us, it might be making the purposeful choice to leave a job that we hate or finding new ways to deal with everyday pressures of life.

GETTING A BETTER RETURN

Again, this is all new information that will take some time to get used to and implement in your life. Although using statistics to reach our goals may sound simple, it must be thought about continuously. It requires determination and concentration.

If, for example, I live in a country in which obesity is a problem statistically, I might want to begin an exercise program to assure myself of remaining fit. Setting my goals would be the first step. I might want to lose a few pounds (most of us do) or get into better shape. Or I might simply want to increase my flexibility, tone my muscles, or keep my heart in good condition.

No matter what my goal is, I must set it, and I must be very specific about how I set it to ensure a better return. I cannot simply begin exercising and expect a specific outcome. Losing a few pounds might require both exercise and a change in my diet. Increasing flexibility might involve training that is different than the kind for losing weight. Toning my muscles or keeping my heart healthy might yet involve a third type of exercise.

THE MONKEY'S PAW

It is important to be specific about goals when setting them. If you want to make $50,000 this year, you would have to be precise about how you would like to make it. We would do well to remember the tale of *The Monkey's Paw* by W.W. Jacobs, in which a couple is given a magical monkey's paw that can grant them three wishes. The problem is that each time they wished for something, they received an unintended result.

The aged sailor who reluctantly gives them the paw explains to them that they should simply throw the magical item from India into the fire, but the couple and their only son, Herbert, do not believe him. The first of their three allotted wishes is for enough money to pay for their house in full – 200 pounds. As the father makes this wish, the monkey's paw writhes in his hand and frightens him, but by the time the small family goes to bed, all has been forgotten, and they fall asleep laughing to themselves about wishes, far-away lands, and magic.

The next day, however, the couple's son, Herbert, is killed in a strange accident at work, and when they are awarded 200 pounds from his employers, they realize what they have done. Later, they wish for their son to return to life – he does, but not as his old self; he comes back without his mind – just a body. What you wish for is not always what you get.

Much like the couple in the tale of the monkey's paw, we have to learn to be specific when we set goals, otherwise our wishes will more likely end in disaster. Therefore, if I want to make sure that I have the best opportunity to make $50,000 this year, I have to be explicit about what I mean. Instead of simply saying to myself that $50,000 is how much I would like to make, I have to also tell myself how I will make this money and within what time period. Will I sell newspapers? How will I get this money? I must also write down my goals in the present tense and in a certain way so that I will be constantly reminded that they will be realized.

In the next several chapters, we will explore particular aspects of your life in which this principle of statistics can be applied. Beginning with chapter five, we will investigate how the power of statistics and choice is relevant to many areas of our lives, including our personal relationships, financial situation, and professional and everyday life. Keeping in mind all that you have read thus far, let us now look into how, specifically, the power of using statistics can help you to make better choices, change the paradigms that are not working for you, and lead you to a better return.

TRULY BEATING THE ODDS: A SURVIVAL STORY

Seven-year-old Jake Paternoster's mother worried about him when he was only five. It wasn't the normal worry of a mother for her son; no, this was something different. She first noticed that there might be something wrong with Jake when he had trouble with headaches, sleeping a good bit, and a decreased appetite. However, one day, his mother noticed that young Jake could not lift a spoonful of peaches into his own mouth. Jake had a brain tumor. His diagnosis was dire; he was not expected to live very long, and the tumor had to be operated on immediately if there was to be any hope at all of saving him. Also, because the surgery was to be so lengthy, they did not expect Jake to walk again. It was even possible that he would be completely paralyzed on the entire right-hand side of his body.

Yet Jake walked again. It took weeks of intensive physical therapy and rehabilitation, but Jake did it. However, the family's joy was short-lived. It turns out that Jake had two tumors originally that were removed successfully, but there was also a third one that had to be removed immediately. This young man had brain surgery twice within a twenty-four hour period. He survived the second brain surgery as well and recovered well. His family says that he never gave up and never stopped believing that he would be well. We should all take a page out of Jake Paternoster's book and learn that, sometimes, beating the odds is just what is meant to happen, or, perhaps, that the odds do not really exist.

Instead, a positive attitude and mindset are of the greatest importance.

CHAPTER FIVE: Effecting Change in Your Personal Relationships

Keep your dreams alive. Understand to achieve anything requires faith and belief in yourself, vision, hard work, determination, and dedication. Remember all things are possible for those who believe.

—Gail Devers

SHAPING THE FUTURE

Our perspective on others is shaped by our past. As life continues, the experience we gain shapes our perceptions of other people. Each thing that we learn about certain individuals will affect what we believe and think about other people.

We discover these things from our parents, our grandparents, and others around us. Learning to imitate the styles and opinions of others comes naturally to us during our formative years, and we cannot help but be affected by those people that we are most in contact with. There are many things we try to emulate from those in our circle, such as what they believe, what they wear, and how they act. And this does not always end with childhood; it extends to our teenage years and, sometimes, into adulthood. It is difficult to go against what our experiences have taught us; it is hard to rebel against such deeply ingrained paradigms.

It is not surprising, therefore, that many of us tend to judge people before we truly know them. I recognize that I tend to put people into categories, or boxes, according to how I perceive their behavior and its consequences, or results. The way that I perceive their actions is mainly based on my past experiences, not only with that individual, but with people in general.

STEREOTYPES

This is where the word *stereotyping* comes into play. Most of us fall victim to stereotyping people at some point in our lives, whether it is for their political beliefs, their manner of dress, or the color of their skin. You have probably heard men and women playing off of their stereotypes. The stereotypical man, for example, is strong, he likes to play football or other sports, and he is the bread-winner and the head of the household. And the stereotypical women are housewives and less capable than men; they enjoy such hobbies such as knitting and cooking.

There are also stereotypes that are related to race. We tend to see others of a different race and judge them, for better or for worse, according to the color of their skin. We all know that stereotyping, for any reason, is not an accurate or acceptable way to judge others, and yet this type of thinking creeps into our subconscious because it is what we have experienced or learned. It is extremely hard to truly escape a particular learned mindset, just as shifting paradigms can be difficult.

If, for instance, you were a student on a university campus who needed assistance with mathematics homework, who would you be more likely to ask for help, the girl who looks like Britney Spears, the guy decked out in preppy clothes, or the girl wearing glasses with a pocket protector in her shirt? What if the "Britney Spears type" might know more about mathematics than the "nerdy" girl? However, you would not immediately think that, because you have been trained by society to stereotype people you encounter.

Here is another example. This one comes from Ninos Malek at the Ludwig von Mises Institute. He suggests a scenario in which you are walking down a very narrow pathway. Your choices are to walk along the left- or the right-hand side of the street. Before making a choice, you notice that on the left-hand side of the street there are about ten men with large muscles, tattoos, and shaved heads, and that on the right there are ten clean-cut guys wearing nice clothing and carrying Bibles. Which side of the street would you choose to walk on?

Whichever side you choose, you would be falling into the trap of stereotyping. The idea is that the ten tattooed, muscular men with shaved heads must be scary or dangerous individuals, but that the well-dressed, clean-cut men must be less dangerous because they look more "normal" and less likely to harm a person. However, these assumptions would be just that – assumptions about these people based on the way they are clad. There is no real way to know which group is the best for you to walk next to; you can only assume that the more normally dressed gentlemen will be less dangerous than the men with tattoos and shaved heads. And your assumptions may be wrong.

People stereotype others on a daily basis without a second thought. We choose to cling to what the norms or society tells us is correct or safe or acceptable because we have been taught to follow certain societal principles. This kind of attitude is most apparent in the way that people in the United States have reacted to Arab Americans after the September 11 attacks in New York City.

A friend of mine was working at a women's resource center on the campus of a university in the United States when the terrorist attacks of 2001 occurred. Some time after the event, the center decided to invite Muslim women to a special gathering on campus so that they could speak freely about their experiences after September 11. The non-Muslim women who attended the meeting were shocked to find that Muslims were having a very difficult time on campus. For example, other students threw things at them and told them to "go back where they came from." They were subjected to death threats and numerous other indecencies simply because they were easily identifiable as Muslims from their dress.

These women and their families were not dangerous terrorists. They were merely Muslims who were either from the United States or had travelled there in order to further their education. There was no reason for them to be treated so badly by their fellow students. However, because a terrorist group of a similar religious persuasion had managed to commit acts of terrorism upon the United States, these Muslim women were grouped with them. The same thing (to a lesser degree) still occurs in the United States today, seven years later.

Many U.S. citizens still get a little nervous when they see someone of Arabic ethnicity sitting next to them on a plane or walking through the airport near them. Even though they know that Arabic people could be from any number of countries and believe in any number of religions, they still fall for stereotypes.

STEREOTYPING AND THE "CAPITAL CITY KILLER"

One strange instance where stereotyping did not lead to accurate information is in the case of the "Capital City Killer." This man, later identified as Mr Derek Todd Lee of St Francisville, was in the habit of murdering young women in and around the city of Baton Rouge, Louisiana. However, it was a long while before he was caught, because the profile that was laid out was absolutely incorrect.

It seems that most serial killers have a pattern. This much is obvious to anyone who has read a story about any mass murderer or serial killer. In fact, recognizing that pattern is how most of these criminals are caught. Yet this serial killer did not fit the profile of any other known serial killer. His first victims were mainly Caucasian women with dark hair. It seems that, according to past law-enforcement experience, serial killers do not ever kill outside their race.

After a while, the Federal Bureau of Investigation (FBI) became involved in the case. Their profile expert was given the daunting task of describing what the serial killer might look like in order to assist police in finding the killer more quickly. The description read something like this: white male, early twenties to late forties. This profile was based upon the race of the women that had been murdered up to that point and upon other available information.

The FBI and the citizens of Baton Rouge were at a loss when the killer was finally apprehended. He was not white at all. He was, in fact, African

American and did not fit the rest of their precious profile either. He was an anomaly, a serial killer that crossed racial boundaries. The inaccurate profile nearly prevented the authorities from catching him and ending his murderous spree in the capital city.

STEREOTYPING YOURSELF

The discovery of truth is prevented more effectively, not by the false appearance things present and which mislead into error, not directly by weakness of the reasoning powers, but by preconceived opinion, by prejudice.

—Arthur Schopenhauer

Just as we stereotype others, we often apply stereotypes on ourselves as well. Your own identity can spell success or failure for you if you allow the way that others see you to affect how you see yourself. Think of all the things that you may have prevented yourself from doing because you did not believe that a person such as yourself could ever accomplish those goals or do such things. What if we all thought of ourselves in a different way? What if we thought that anything was possible, that stereotypes do not apply to us, just as they don't apply to others?

We must accept ourselves for who we truly are and learn to love ourselves in a more meaningful way. Limitations upon our abilities are too often self-imposed. Perhaps we believe others when they say that we are not smart enough or do not come from the proper background to be successful. Push these thoughts from your mind. Everyone is beautiful and unique in his or her own way and has talents to share with others. The only one who can keep you from seeing your inner worth is yourself.

We have been conditioned to think that we should look or act a certain way in order to be accepted by society. This is simply a way of self-limiting ourselves. Anyone can accomplish everything that he or she wants. It is a simple matter of believing in yourself and not allowing stereotypes to affect your own self-image. We can all understand the concept that what we think comes into being. So if we think of ourselves as unsuccessful or unhappy, then no matter what the circumstances are, we will remain so. I read somewhere that Napoleon said that he always created his own circumstances, that he did not allow circumstances to rule his actions. This is how we, too, should be.

People are not simply made up of their experiences and their past. People cannot be judged or defined by the clothes they wear, the neighborhood they are from, the family they were reared in, the car they drive, or the color of their skin. We have been taught that these accidents of birth are somehow important and that they should define us. The truth of the matter is that we are the ones who define ourselves.

If we believe negative things about ourselves, then those negative things will become important because we have made them important. On the other hand, if we keep a positive mindset and think good things about ourselves, then those things will be true as well because we have made them so by believing in them. Our self-image, the perception that we have about our own characteristics, is a powerful thing. It can mean that we could become successful, happy individuals who are not afraid to take action. Or it can mean that we limit ourselves to living a life that is not abundant.

Each person in this world has a powerful and indestructible force within himself or herself – the power to think negative or positive things into being simply by believing that they will happen. If you change your mindset to believing wholeheartedly that you are a positive, beautiful person, then you will become that person. In fact, you already are a beautiful and strong person; you simply have to make yourself remember how wonderful you are.

Avoid creating negative stereotypes of yourself. Do not ever think that you are limited by what society tells you is correct or incorrect; you are the one who is in control of your own life and destiny. Instead, take action. Begin believing that you are powerful and that nothing can stop you from achieving your dreams. Visualize yourself being successful and you will become successful. That is one of the secrets of successful people: they never allow anyone to tell them that they are wrong. Instead, they take action toward making their dreams a reality.

Many people before me have suggested that you should believe in yourself. Why not take it seriously this time and truly work at it? Nothing will stop you except your own self-image, which is a powerful thing. It can truly mean the difference between success and failure in everything you do. A negative self-image can be related to fear and anxiety about your life and about the future.

However, there is nothing to be afraid of. Your dreams are simply waiting for you to gather the courage to realize them. You already have the courage within you to make these decisions and to become the self-confident person that you would like to be. If there is something that you do not like about yourself, simply change it. If you are less kind, for example, than you want to be, then just become a kinder individual. Work at it and persevere until you become the person that you want to be. Nothing, except your own mindset and your subconscious beliefs about yourself will assist or hinder you in this endeavor.

You can become the person that you want to be, a person who is happy and fulfilled in every aspect of life. Choosing to live that life is the first step in the process. Choose the person that you would like to be and then work hard at becoming that person. Even Aristotle used to say that virtues can be learned by acting virtuous. By the same token, you can learn to be the person you want to be by simply becoming that person. The results you want will come if you take action toward them.

THE CHRISTMAS TRUCE – WHEN STEREOTYPES ARE BROKEN

What we seem to often forget is that, for each person, circumstances are different. What may be right for me might not be right for another person. Something that is acceptable to me may not be acceptable to others. Take, for example, the Christmas truce of World War I.

On Christmas Day 1914, in a small village near Ypres, Belgium, an amazing thing occurred. British and German troops suddenly found themselves unable to fight one another. Instead, they shared Christmas gifts, meals, and general festivities. It was, perhaps, the most famous unofficial truce ever recorded.

These were soldiers that were supposed to be mortal enemies – literally. They were at war, with each side dug in on the opposite sides of the trench that separated them. Mortar shells littered the area between them, and the dead bodies of their fellow troops were left there as well. I suppose it was the Christmas spirit that caused the truce, but when the bells in the nearby town rang, neither side could fire another shot.

All throughout the day, they ate together, played games, and exchanged hats, buttons, and other small tokens. By the end of their Christmas meal, they wondered how in the world they had ever been shooting at each other in the first place. Legend has it that both sides had to send in replacement soldiers because they refused to fight each other any longer, having gotten to know one another.

This is a perfect example of putting people into wrong categories. When a country is at war, there is no thought given to anything but victory. What is forgotten is that people, not countries, fight wars. The British troops most likely hated the Germans with all their hearts, as did the Germans hate the British. However, when it was all said and done, they were able to get along quite well and get to know one another not as enemies, but as friends during a time of celebration that is supposed to symbolize world peace, not world wars.

IN REAL LIFE

Yet being aligned with one country or another is not the only thing that can breed contempt for fellow human beings. If I see a person who is dressed in a particular way, I might think certain things about that individual. I might think that person is rich, poor, mean, easy to get along with, or unintelligent. All of these perceptions and prejudices would be based on my experience with other people who dressed or looked like that person. This does not mean that my perceptions would be correct.

This is simply a way to give power to the stereotypes we discussed. If, for instance, I am a particularly conservative person, I might be more inclined to look negatively upon a person who has tattoos or strangely colored hair. If I am a person with tattoos and strangely colored hair, I might be more likely to look down upon a person who dresses conservatively. None of these outward qualities really tell me who a person truly is on the inside. It is all a matter of appearances, and outward appearance will not guarantee good or bad results.

Why I put other people into categories does not always have to be related to the way that they are dressed. People may assume things about others by the way that they act. By the same token, I could fall into the trap of prejudging people, or creating stereotypes, by what kinds of results they get in their lives, whether in business or in personal situations.

I know a man who is what I like to call a master planner. Most of us probably know a coworker or friend who is like this person. He spends most of his time *planning* what he will do instead of *taking action*. If he were to use one-tenth of the time he spends planning on actually doing work, he would be the best in his field. Often, I am tempted to judge him by the results he gets from his work. And the results he gets are not the best; I am, therefore, tempted to place him in the category of people who are not successful. However, recently, he asked me to assist and coach him to become more successful, so I have had to reconsider the box that I had put him in.

Clearly, the theories that I use to put people into boxes, which are inherently confining, are often simply wrong. Instead of putting people into categories, we should get to know people before we decide what box they fit into. People deserve to be treated as individual human beings, with their own distinct personalities and character traits, and they should only be judged after we know them a little better. We must give them a fair chance before jumping to any conclusions.

In order to accomplish this, our sense of awareness has to be heightened. Learning to become a better listener and understanding how to truly know people are valuable skills, more so than most of us realize. There is an old saying: we have two ears and one mouth. This means that it is good to listen twice as much as we talk. Learning, for example, can only happen when we are listening to other people as opposed to talking to them.

Here is one example of putting people in the wrong box. I have a friend who is a professor. When I use the word *professor*, many people tend to think of an older man with a beard and a tweed suit. However, my friend is a young woman who dresses in a much more modern way. She does not fit the professorial stereotype, and yet she is indeed a professor. In fact, her students generally respect her more for not fitting the stereotype. They are pleasantly surprised on the first day of classes to see someone other than a cold and analytic professor in a stuffy suit.

The idea is that we should never put people into the boxes. Some of us are more private than others, and it takes a long time for others to get to know us. Others of us are more open, so it takes less time for other people to get to know us. When we put others into categories, however, we miss the opportunity to get to know the "real" person inside. Simply because people dress or act in certain ways does not mean that they *are* that way.

CHANGING YOUR RELATIONSHIPS THROUGH CONSCIOUS CHOICES

I have never looked at the consequences of missing a bit shot... when you think about the consequences you always think of a negative result.

—Michael Jordan

There are several ways in which you can change your relationships with other people. One of these is through making conscious choices. The first thing to do is to become more aware of what each relationship means to me. I have to ask myself: What does this person's presence in my life represent? How much does the person mean to me and why?

It is not unusual for a person to have different kinds of friends – from those that are fun to be around to those that are always there in times of need. Determining how important these friendships are will affect how and in which direction the friendship develops.

Some of us have forgotten what true friendship is. As we get older, real friends become rare, and we begin to cherish the ones we have managed to keep and to cultivate. However, this understanding also makes us realize what a real friend is. Deciding whether or not a friendship deserves my utmost attention is difficult. Not everyone that I would like to be friends with will be someone I would like to cultivate a meaningful and lasting friendship with.

Is this person the type of friend that will help me when I need assistance? Will the person be willing to help me if I move across town, or be there for me when I need someone to talk to or a place to stay?

A coworker of mine once had a twenty-year friendship with someone. Each time that person visited him, he wondered why they were friends. My friend got no benefit whatsoever from this friendship. He did not receive understanding, consideration, or assistance with his problems. Also, he felt that he did not have more in common with this friend after such a long friendship. They had grown apart, he was sure, but he couldn't put his proverbial finger on the reason why.

Finally, my friend was able to describe the situation to me. The two friends, being different now than when their friendship began, had an argument the previous night. They were unable to understand each other's differences, which forced my friend to take account of the relationship. His friend said, "But I don't understand. We have been friends for twenty years!" My friend replied, "Have we been friends for twenty years, or were we friends twenty years ago?"

Seemingly, these are harsh words, but they are also true in some circumstances. Oftentimes, we must rid ourselves of the negative influences upon our lives, even if that means rearranging some of our friendships and other personal relationships. We certainly cannot change and grow as people if there are others who do not wish to change and grow around us. Good friends, even old friends, will want to support you in your new endeavors; they will not be interested in tearing you down. Instead, they will concern themselves with lifting you up in this time of change and growth.

If I know which friendships will be worth my while, I can determine how much energy I will expend on fostering the relationship. I can decide which direction the friendship will go in. I can determine how much time I want to spend with different friends, how much I will do for them, or how much of my personal affairs I will share with them.

However, the most important question that I can ask when it comes to any personal relationship is what I can do for the other person rather than what that person can do for me. This is the most unselfish choice, and it goes back to the idea that we can learn more from people by listening instead of talking to them; in so doing, we are able think about what we can do for others instead of what they can do for us.

Perhaps the better questions to ask are: Will I be there when this friend needs me? Would I do all of the things that I expect a good friend to do? Will I be there for the person during the tough times? When we begin to ask these questions, we really begin to touch upon the point of making conscious choices in our personal relationships: that choosing to change ourselves will effect change in others.

CHOOSING TO CHANGE YOURSELF CHANGES OTHERS

Throughout our lives, we have been taught to look for what is wrong with things and situations instead of what is right. In school, our teachers used their red pencils to tell us what was wrong with our work. Mostly, we were taught what was wrong with our results or behavior, whether at home by our families or at work by our bosses.

This kind of thinking affects how we treat others. I often catch myself finding fault in others because I am accustomed to finding fault in so many other things; this is part of what I have learned from the paradigms that have been placed upon me since birth. It is also because I have been taught to find fault in myself; as such, I tend to turn that frustration outward onto the others in my life, finding fault with them instead of looking inward at my own issues.

When we find fault in others, it is usually indicative of the fact that we are unhappy with ourselves. Often, we find the most fault in people that are most like us. However, we must move away from this habit and turn our gaze inward. Taking action to change our own small flaws will allow us to understand others better and to have more meaningful relationships with them.

Once you begin to see that changing yourself will not only change your perception of yourself, but also the perception that others have of you, you will understand how important mindset truly is. Having a negative self-image and mindset will create that illusion for others as well as ourselves. This situation that is easily avoidable and simple to remedy will, in turn, create havoc in our lives.

There are several tools that you can use to accomplish this task. The most important thing that you can do is to take a look at yourself and your habits and take action toward changing them. Another thing that you can do is to teach yourself, through practice, to perceive others in a more positive light. Once you change your perceptions, the world and the people around you will take on a new and powerful meaning. You will begin to see everything as a positive force. Even the things that you first believe are failures or obstacles will now be perceived as things to be thankful for; they become stepping stones instead of stumbling blocks.

GRANDMOTHER'S LAW

Grandmother's Law instructs us to always look for something good in other people *and* make sure you tell them about it. I call this "Grandmother's Law" because this is typically how my grandmother, Borghild, thought and acted. She always praised us for the good things she saw in us, and she never had a bad word to say about anybody. My grandmother was able to perceive others as they should be perceived – as people who are worthwhile, no matter what their faults may be.

When you apply Grandmother's Law, however, you must believe in it. If you apply this law in a careless or insincere way, it will come across as criticism instead of praise. Making others feel good about themselves is an example of how changing yourself will change others. They will appreciate your new-found or amplified positive disposition, and they will desire to be more positive themselves.

If we do this, the people we come into contact with will remember this kindness much better than if we criticize them. If they have done something wrong, they best know what they have done. They also know why they did it and, therefore, do not need you to rub it in. If you tell them good things, it will make them feel good; they will respect you, and your relationship will be stronger than one based on criticism and negativity.

For instance, if there is a coworker who is very difficult to get along with, arguing with him or her will hardly ever resolve the problem or conflict in question. Instead of finding fault with everything that that coworker does, try telling the person about what *is* working instead of what is *not*. This will foster a better atmosphere in the workplace, and this will benefit everyone who works there.

This is also a good rule or law to remember if you are a supervisor or in a managerial position. If you constantly find fault with others, the people you work with will see you as a person who is difficult to work with and who is always complaining. This may not be the reality, but oftentimes perception is reality. If you, instead, apply Grandmother's Law in regard to your employees, you will get better results in the workplace in the long run.

APPLE POLICY

Steve Jobs, cofounder of the Apple corporation that brought the personal computer to the world, has an interesting policy for his employees. He determined a long time ago that people are more productive and happier in the workplace if some of the traditional rules are thrown out of the window. The employees who work for Apple are not, for example, chained to any particular dress code or way of working. Free thought and less corporate control define this management style.

And the policy paid off. Apple computers are highly innovative and sell quite well. This is a real-life version of Grandmother's Law in action. Employees are reminded every day that their boss expects the best from them, but is also willing to give them the leeway to produce products and ideas as they wish.

YOUR ROMANTIC RELATIONSHIPS

In romantic relationships, the same rules apply. If you can always find something good to say about your partner, then the lines of communication will remain open, and your partner will give you back the same respect. Arguments can be diffused if a simple, open-minded, and positive approach is taken from the very beginning.

You probably know a couple, or are part of one, who enter into arguments frequently. These disputes are hardly ever about the things that are being discussed. Each partner reasons that many small things turn into large arguments because the other has a preconceived impression of him or her that colors the issue. We react differently to the comments of people that we have close relationships with than we would to a stranger's similar comments.

One reason for this is that we believe that we know our partners or close friends so well that we can predict what they will say. We then react to that prediction without giving the other person any opportunity to respond at all because we assume we know what they are going to say. The other reason is that we tend to treat the ones we love a little differently simply because we know that their love is supposed to be unconditional. You have heard the old saying, "You always hurt the ones you love," which is true in this case. We are less afraid to anger people we truly love because we know that they will still love us.

Learning to find the positive characteristics in your partner is of great importance. When people are in relationships for extended periods of time, as I have explained, they believe that they know their partner so well that they can predict how the partner will respond in most situations. I maintain, however, that this is not true. Your partner is still a human being who is completely unpredictable. What your partner has done in the past is meaningless. It is, however, important that you see your partner for who he or she is in each and every moment. Use Grandmother's Law and think of all of the positive and wonderful things that made you first fall in love with your partner.

Thinking of the good things about your partner instead of the negative, or bad, things will make your relationship much more smooth and easy. Not only will communication become less difficult, but your partner will appreciate your concern for their feelings and respect for them as a person. This is invaluable advice for anyone looking to remain in a long-term relationship.

Being positive about all aspects of your life, especially your personal relationships, can create a positive attitude in others. When you go to your partner with a positive mindset, he or she will be far more likely to listen to what you have to say than if you were to attack with harsh words and criticism. Finding the good things about other people will make your life easier and more fulfilling; it will also enrich and contribute to the growth of your personal relationship.

Statistics about relationships can be scary. There are only a small percentage of people who are in steady and truly happy relationships. However, we can use that information to avoid certain pitfalls. Fostering a better atmosphere is not only for the workplace. In romantic relationships, we should create an atmosphere that is loving and giving, instead of waiting for our partners to give to us. We should, once again, think of what we can do for our partners instead of what they can do for us. Learning to give as often as possible is very important.

One issue that can become important in friendships, romantic relationships, and other types of relationships is the past experience that we have had with the other person in the relationship. When we first meet someone, we do not have a history with that person. However, after that, we quickly place that person in a box and are rarely open-minded about him or her. After they have been placed into categories in our mind, everything they do and say seems to fit what we already believe about them.

If you have been friends with any one person for any length of time, you will notice that often the same subject matter keeps coming up, even when you are having a disagreement about something entirely different.

If the disagreement begins as an argument about where to go for lunch or a vacation, you both suddenly find yourselves in the middle of a conversation with phrases such as, "You *always* act this way when ..." or "You have *always* taken things so personally." Another example: "You *never* allow me to finish a sentence," and on and on.

This occurs because these two friends have placed each other in their own respective boxes (even though they know each other well), and also because they have not chosen to change in order to effect change in others. If they were to choose to make every encounter with the other a learning and listening experience, then both friends would be satisfied with the relationship. An atmosphere of mutual respect and admiration would develop, and both parties would be better for it. Making choices that allow your relationships to grow will always affect that relationship in a positive way, proving that statistics can be changed or beaten.

Choosing to change yourself and how you deal with others, in effect, changes them. Be aware that pointing out the positive aspects of others instead of finding fault with them will change how people see you; it will also change how they, themselves, act in subsequent encounters with you. Here are several more examples.

LITTLE PITCHERS

You have probably heard parents or grandparents say that "little pitchers have big ears." This old saying is about how children not only hear everything, but also soak in what they learn from the people around them. And this is how our environments really affect us. Almost everything we know is directly related to our experiences, just like children who take in everything that they hear and see and internalize it. Knowing this, we can choose to change the way we behave and, thereby, affect others.

This is not unrelated to the way that we change our behavior around children so that they do not pick up our bad habits. We want the children we are responsible for to act in a certain, acceptable way, and, therefore, we are careful not to pass our negative behaviors on to them. Being a role model is not always easy, but we manage to do a pretty good job. Why do we do this? Because we know that whatever we do, our children will mimic us. If we teach them how to become the people we want them to become when they grow up, then we have affected their lives by making conscious choices to do so. We choose to be their role models because we know that what we do affects their "results."

In the same way, if children are treated as if they are stupid, ugly, or worthless, then they will feel that they are this way even though they are not. How we treat people gives them an impression of themselves. Their mindset becomes crafted by the language those around them use about them. This is even more important than in romantic relationships because children are more likely to believe what others say about them than they are to be self-confident.

CYBER BULLYING: LANGUAGE AND MEANING

In the last several years, the problem of "cyber bullying" has become a huge issue. What most teens and others would not say in person, they feel comfortable typing and displaying on the Internet. I once heard a story about this very use of the Internet for cyber bullying. Some students in middle school created a website where boys could rate the way that girls looked. Many of the things that were written were untrue, but most of the boys thought it was a funny joke. What they did not realize was that the girls were extremely hurt. The next day at school, several of the girls who had been written about were in tears, including one boy's little sister.

The point is that the language that we use can be hurtful, even when it is on paper or displayed over the Internet. The words that we use to describe people and to talk about them must be selected carefully. We might end up hurting other people's feelings or causing them great distress. These kids had no idea that their words about their schoolmates would ever be seen by anyone else, but it qualifies as bullying anyway.

Even as adults, we tend to do this. We do not choose our words or our attitudes carefully. It is tempting to go into a coworker's office to gossip or to complain about others in the workplace, but that does not mean that it is ethical. Language can be just as hurtful as actions. Some would argue that language or speech can be more hurtful than actions because it is difficult to forget the things that are said about us. We tend to remember the hurtful things and forget the positive things.

We must avoid a negative mindset of any kind. When we bully others, we hurt ourselves more. Using negative words or actions in place of positive ones will never resolve any conflict or solve any problem. Whether it is cyber bullying or negative language, we and others are hurt by these harmful attitudes and words. This is especially true, as we have discussed, with children. However, here is a more personal story from my life to illustrate the point.

MARA'S STORY

I once knew a girl named Mara; she was from Somalia, she was sixteen years old, and she went to school with the daughter of a friend of mine. Mara had some serious problems at home. She drank, her behavior was out of control, and she was generally headed on a path of destruction.

One night, my friend's daughter called and asked me to pick Mara up right outside of the city, where I found her, sitting surrounded by garbage bags that contained her clothes. Mara's mother had finally had enough and kicked her out of the house after calling the police. I took Mara home, and for about one year she stayed most of the time with me in my apartment. It was a very turbulent year.

As I got to know Mara a little better, I realized that one of her biggest problems was with self-image; it was one of the most negative I had seen. She had been affected by the way her parents treated her, and she had begun to believe all of the terrible things they said about her. She got her negative self-image from them and from other people around her.

I thought for a long while about how to help Mara to see herself for the beautiful and promising young woman that she was and is. Finally, I came up with an idea. It was very simple; I bought a pad of small yellow sticky notes and wrote wonderful things about Mara on them. I left each one in a different place: one on her pillow that said "You are a wonderful girl"; another on the bathroom mirror said "You can do anything!" I made sure to write nothing but encouraging words.

You can imagine, then, that I was not surprised when Mara began to respond beautifully to this encouragement. I was able to help her change the negative view she had of herself with these tiny notes. Her self-image did not change in a month or even a year, but it was a good beginning. Mara had her ups and downs over the next three years; she slid back into some old, negative behaviors, but I continued to help her.

In the end, I am happy to report that Mara is doing very well. She now lives with her daughter Aisha in a nice flat, and she studying to be a nurse as she had always wanted to be. Most importantly, she is happy. She always makes sure to thank me when we speak; she tells me that she would not be alive if I had not picked her up that night that her mother threw her out. But I just tell her that there is no telling what would have happened if I had not picked her up; it's all in the past now.

AN EXERCISE

There is a common exercise that will allow you to see how the way we treat people affects others. Once, my professor wanted us to do an "ice-breaker" exercise on discrimination. She asked us all to pick numbers (ranging from 1–10) out of a hat, but we were not allowed to look at them. We were simply instructed to place these numbers on our foreheads, so that others could see them, but we could not. The professor explained that we should walk around the room and treat our fellow classmates according to the numbers. If a person had number one, he would be treated the worst, and if the person had number ten, he would be treated very well. The object was to guess which number was on our foreheads without cheating.

It was an eye-opening experience. I thought it would be difficult to determine which number I was wearing, but it was immediately apparent. The way that the other students treated me affected my perception of myself, if only accurately enough to guess which number I wore. Yet this is how real life is. Some of us are treated as a "one" and others as a "ten" for various and sometimes random reasons. If you ever find yourself in a position to try this experiment, it is worth participating in. It shows, also, how easily we put others in boxes because of their actions and their outward appearance. In addition, this exercise can teach us about how it feels to be treated differently than we normally do and to learn how much it really does affect our vision of ourselves.

APPLYING GRANDMOTHER'S LAW

Many years ago, I got a new job as a manager for the western part of the country in one of the biggest transportation firms in Norway. Before I even arrived, I was told that the people and the situation there would be very difficult to handle. I thought to myself, "What should I do in this situation? How shall I handle this problem?"

So, as part of my preparation for this new job, I began to wonder how I would deal with these people who were supposed to be challenging. My thinking is that people are people, wherever they live. It does not matter whether they are in the western portion of Norway or on the moon; they will still have the same basic needs and react the same way in most situations. I decided that I would put Grandmother's Law to work.

I came up with a great plan. Every week, I decided, I would tell all of my employees something positive about their work, instead of pointing out the negative. I wanted to see whether or not it would actually change their behavior and attitude. Did my theories work?

The result, needless to say, was overwhelming. Not only did my theory work, it worked wonders. In all areas of the firm, we witnessed tremendous growth, and everyone, to their surprise, seemed much happier. The workplace was a much better environment than when I first arrived, and Grandmother's Law had worked. When you tell people positive things about themselves, you must always be sincere; otherwise, you will run into the danger of effecting the opposite kind of change than the one you intended.

For instance, if you were to walk into a classroom full of students that you had already been told are "bad" by someone else, you might be tempted to treat them as such. However, if you apply Grandmother's Law to this situation and encourage these students, instead of criticizing them, then you will get better results from those students.

Once, I had a grade-school teacher who treated all of her students as if they were absolutely dumb. Even the good students were constantly ridiculed and told that they were stupid. When a student failed a test, this teacher would hold it up in front of the class just so she could make fun of that person. With all of her attempts to shame students into behaving and into getting better marks, she never succeeded. None of us were ever able to get past the fact that we were terrified of attracting her wrath to get anything done, much less to improve.

However, I have also had teachers who were quite the opposite. They were comforting when things were not going well; they were tough, but fair. They made it a point to reward good behavior and to foster an atmosphere of mutual support in the classroom. Not all of us were gifted students, but because some teachers treated us that way, we felt as if we were and, in fact, achieved more and got better results.

CHAPTER SIX: Effecting Change in Your Financial Situation

Nothing in this world can take the place of persistence. Talent will not; nothing is more common than unsuccessful people with talent. Genius will not; unrewarded genius is almost a proverb. Education will not; the world is full of educated derelicts. Persistence and determination alone are omnipotent. The slogan 'press on' has solved and always will solve the problems of the human race.

—Calvin Coolidge

MONEY MATTERS

Your personal relationships are only the first of many areas of your life in which you can use the power of statistics to get better results. Perhaps one of the most important areas is your financial situation. Let us apply the principles we have discussed thus far to money matters.

Finances are something that all of us worry about. We wonder what the best way to handle our money is, how to budget, and how to best plan for the future. Yet most of us barely understand what the best pathway to the results we want is. More often than not, we try to manage our money in a way that works for us, but we still end up with the same unsatisfying results.

Remember that the definition of insanity is to keep doing the same thing over and over and expecting different results. No one should expect different results if he or she continues to do things the same way. If I burn my hand every time I touch a hot stove, then I should notice that perhaps touching the stove is not a wise move in the first place. The same goes for your life; if you keep doing the same things, you will continue to get the same results. Repeating bad habits and remaining in the same confining paradigm, especially in financial matters, can end in disaster.

YOUR MINDSET MEANS EVERYTHING

It all comes down to your mindset. Mindset is very important in all aspects of your life, but it is perhaps even more important when it comes to your income. If you think you cannot do something, then you will assuredly fail to do it. However, if you are able to change your thinking, then you will succeed. In order to change your thinking, as we have already discussed throughout the last several chapters, you must also change your actions.

If you change your actions according to your new goals, you will get new, and better, results. What is most important to remember is that the past is just that – already gone. Things that have already happened cannot be changed, and we should not, therefore, worry about our past actions. We should, instead, look toward the future. Changing our actions and habits, our paradigms, is beneficial. It all comes down to how we think of ourselves. Just like the story of Mara, it is our self-image and what we think of our abilities that control what happens in our lives.

This is especially evident when it comes to finances. Many of us believe that we cannot change our situation financially; we live with a mindset of scarcity. Few of us understand that we possess the power to change this mindset into one that leads to a life of abundance. When we live with the mindset of scarcity, it is difficult to see the world any other way. Getting stuck in this way is common, but the situation can be changed.

For instance, many of us have different financial responsibilities. We have bills to pay, mortgages to worry about, and we all want to make more money than we already do. However, if we keep ourselves from living in abundance, we will live in scarcity. This means that we act and make decisions based on our past lack of abundance and believe that it will be too difficult to change things. It is easy to remain in this paradigm because of all the little things that happen every day.

These days, saving money is not enough. Almost all people are concerned about their success level, as well as their future. Will you be able to retire, to travel, or to do any of the things that oftentimes seem out of reach? Yet what does "abundance" mean?

LAURA'S CHOICES

The concept of abundance depends on what kind of surroundings one is used to. As such, abundance could mean more money, true love, lasting relationships, or any other thing that is important to you. Even though no one would argue that money is not important, it is also not the most important thing for everyone. And the amazing thing about abundance is that you can choose it, just as many of us have chosen scarcity without realizing it.

There is a woman that I know – let's call her Laura – who cannot be convinced to live under any other paradigm than one of scarcity. She is a person who has a good job which more than pays for her bills, yet she still has trouble every month managing the money that she does make. The payments for Laura's house are not unreasonable, and her other expenses do not amount to much, yet she is always complaining to me about how she cannot afford anything that she wants. This is because she lives with the mindset of scarcity. In fact, a friend of ours said that Laura will never have the money she wants because she will always believe that she has too little, no matter how much she actually accumulates. This is not to say that she will never be satisfied with any amount of money, but that she will always believe that money is scarce, no matter how much she has. Laura's state of mind is one of scarcity.

You see, this woman did not grow up with many things. She and her family lived under the paradigm of scarcity. She remembers that, growing up, her family hardly ever had the things that they wanted. They had what they needed, but nothing was ever left over. Now that she has grown up and has a better potential for future earnings than her parents did (she was the first to graduate with a degree from university), she still has the same frame of mind that she had as a child.

For example, she spends more money than she should on buying items in bulk that eventually get wasted. As such, she is more likely to purchase cheap foods than to buy enough healthy food for herself and her family.

This woman has never understood that it is better to buy just what you need than to waste money on so-called bargain products that are not as good for your health.

What Laura fails to understand is that scarcity is something that she chooses, just as she cannot understand that abundance is something she can choose as well. In other words, past financial mistakes or environments do not determine your potential. Your financial future is an unwritten page, and you can change it now, right now, by choosing to develop an attitude of abundance.

POSSIBILITIES AND POTENTIAL

Think of the potential within each one of us. The possibilities are endless, because our capacity to think is infinite. What has happened in the past, financially, is no longer important once you operate from the mindset of abundance. Only the future matters, and the future is full of endless possibilities. People live with the mindset of scarcity because it is a result of past ways of thinking. Like Laura, we become accustomed to these ways of thinking and find it hard to break free, even if these modes of thought are not working for us. In fact, to get results, don't think – act.

Taking action is a key part to living with the mindset of abundance. Laura, for instance, was once offered a substantial raise at her place of employment. However, she turned it down, because she was afraid that she could not handle the additional responsibilities. Laura, once again, chose scarcity over abundance and, most importantly, did not take any action at all. She could not control her financial future, even in the face of a wonderful opportunity, because she did not believe in herself. Instead, she chose to live in scarcity.

In fact, choosing the correct statistical option in the present ensures your financial future. We all know that many people carry a great deal of credit card debt, statistically speaking. Again, the same principle applies. Using those statistics to inform our decisions can affect the outcome of our financial lives. We can make the choice to control spending, save money for purchases instead of putting them on credit, and thereby beat the statistics and begin to understand what living with the mindset of abundance really means. Now, let us move on to how you can live in abundance by taking action in your chosen career path.

SUPPLY AND DEMAND

Some questions that you might be asking yourself: How do I take action? How can I change things and choose abundance? Let's first look at how success and job performance are determined. Practically speaking, there are three things that determine a person's income: (1) the demand for the job you have; (2) how good you are at doing what you do; and (3) how easy or difficult it is to replace you.

First, there must be a demand for what you do. This follows the well-known economic rule of supply and demand. If there is no demand for the product you are trying to sell, namely yourself, then it is nearly impossible to succeed. For example, many successful people notice a need in their market and they step forward with the "supply." You probably already know about some of these ideas; perhaps the most obvious is the demand for weight-loss products.

We see them everywhere: diet pills, weight-loss programs, low-calorie sodas and foods, and various other goods designed to help people to lose weight. Oprah's weight is the subject of grocery store checkout line gossip, and images of young, thin women are considered to represent the pinnacle of beauty. Just about everyone wants to look beautiful, to be accepted, and to be loved. Many people feel that a perfect body is necessary for happiness. Businesses take advantage of this fact and meet that demand with the necessary supply: weight-loss products. Some of them work and some of them don't, but there is definitely a market out there for helping people to feel better about themselves. Most people do not understand that true acceptance comes from within, because they have not been taught that by society and their environment, so there will always be a demand for these types of goods.

Yet a demand for what you do is different from the demand for certain products. What is in question here is whether or not your job is actually in demand. For example, in the United States during the 1990s there was a huge demand for anyone working on websites or the Internet. Since then, the demand has diminished, but in the heydays of the Silicon Valley, these occupations were highly in demand.

I have a friend who worked his way up the corporate ladder, making himself useful and doing his job quite well. After years of this, the company decided to make some cutbacks. My friend could no longer make a living while remaining employed with that group. However, once he left, everyone around him realized just how much work he really did around the office. It wasn't too long before that same company began to miss him and his hard work. He created a demand for the type of employee he wanted to be. They invited him back to his original job for a higher salary, and he thus became even more successful at his job than before. In other words, I must make sure that there is a demand for what I do. If not, it would be better to find a new area to work in.

BEING GOOD AT WHAT YOU DO

The second thing that people get paid for, besides the demand for their job, is how well they do their job in the first place. Being a hard worker is great, but if you are an inefficient or unreliable employee, then you should not expect to get ahead. However, if you are dependable and do your job well, you will certainly be compensated for that.

There are plenty of people in the workforce who arrive at their job day after day and simply do what is required of them. Doing nothing more than the minimum is what these employees seem to be interested in. Having this attitude can never make you happy. However, it is not difficult to be good at what you do. First, you must have an attitude of always doing what you do to the best of your ability. Then, if you find that your job is not one that you want to stay in, identify what it is that you love, and then go out and find that job.

Taking action is, again, important. How good are you at what you do for a living? Does it fulfill you and make you happy to go to work each morning? Are you truly taking the time to complete each project that is assigned with great energy and seriousness? If you can do these things, you will be better at your job, and you will get better results. This will also make you feel better, knowing that you have done your best.

Let us say that you do not enjoy your job. Many people work in places that they do not enjoy. Therefore, they are not interested in doing a good job and are not going to be good at it. What are your strengths?

ARE YOU REPLACEABLE?

The third thing that will determine how you are paid and whether or not you are paid well is how easy it is to replace you. Sure, education and hard work can get you somewhere, but what happens when you can be easily replaced? You are more likely to lose your job if there is someone else that can do your work as easily as you can do it. It is important, therefore, to make yourself invaluable to your company and to your boss.

I know this to be true because I have seen it happen quite often. Once, because of pay cutbacks, a friend of mine was forced to seek employment in an area other than what she was trained for. Everyone at her old job was sad to see her go, but they did not think much of it at the time. She had always been a well-liked, hard worker. She never grumbled, she hardly ever said no to any new challenge or responsibility that was given to her, and she went about her work in a happy manner. She also loved what she was doing for a living and was terribly disappointed that she would no longer be able to work in her old position.

Once she left, however, something amazing happened. Her boss tried to replace her, but could not. The person sent in to do her job was quite bad at it and also undependable; in fact, the worker quit in the middle of several ongoing projects. It was only then that her company discovered how lucky they were to have had her in the first place. They asked her to come back, nearly doubled her salary, and offered her several perks that she did not have before. You see, she had made herself irreplaceable. Clients did not want to deal with anyone else but my friend, so the company was forced to admit that they needed her more than she needed them.

This was a unique situation, and no one would suggest that the best way to be noticed is to leave or quit a job in the hopes of your boss realizing how irreplaceable you are. But, while you are working in your current field, you can do some things, like my friend did, to make yourself a valuable employee. And next time there is a raise on the table or an opportunity for advancement, you will be prepared to accept it. Some people call this luck, but it is not. Luck is when preparedness meets opportunity.

So, again, how much you get paid and how much opportunity is afforded to you is based on three things: the demand for what you do, how good you are at doing what you do, and how easy or difficult it is to replace you. What I have to do, based on this idea, is to make sure there is a demand for what I do; then be as good as possible at what I do; and, lastly, make sure I am not easily replaceable.

THE SUBCONSCIOUS

I have recently been reading *The Subconscious Mind* by Joseph Murphy, and this work brings home the point that our subconscious, what we have up until now referred to as a mindset, is of real importance to our lives. Murphy argues that the subconscious mind is not just *an* important thing, but *the only* important thing in our lives. He maintains that everything that we see, or are exposed to in any way, becomes a part of our subconscious. The idea is that whatever you put into your subconscious will affect your results. This works the same was as a mindset. Both are paradigms that one cannot escape easily.

Murphy says that everything – illness, health, wealth, or any variety of things – is affected by our mindset, by our subconscious. It is your strong belief in something that controls everything that you do. If, for example, you truly believe that you will never be successful or wealthy, then you will certainly never be either. If, on the other hand, you believe wholeheartedly that you can do these things, then you will accomplish them.

You see, what we believe about ourselves can make a huge difference in how successful we are, especially in financial matters. It sounds rather strange to talk about success and wealth in this way, because this is not the way that you are accustomed to thinking about money. Most of us are taught that making money is a very tricky thing; we learn this throughout our lives – making money is hard, and it cannot be earned without cheating or swindling others. We see that many rich and successful people have had to step on and hurt many people to get to where they are. This is where the phrase "It's lonely at the top" comes from; we assume that anyone who is wealthy and successful must have done something unsavory in order to reach their goals.

However, this is related to a belief system, a mindset, that has been burned into our subconscious by our various experiences thus far. Yet this is the very mindset that I wish to debunk. You can choose to think differently and, in making that choice, change what your subconscious tells you. This is what it is all about – changing your thinking and then taking action accordingly.

We must think of success and wealth as something that is absolutely attainable, not something that is hard to get or that will never happen. Instead, think of living in abundance as if you already do. If you do this, your results will be better, and you will find yourself doing things that you only ever imagined you could do. You will live with this mindset because you will have changed your thinking.

Clearly, it all depends upon what you choose to expose your subconscious mind to. For example, what if we were discussing the best way to overcome stage fright when speaking in public? Giving a speech in front of a large number of people can be terrifying for many people. What kind of qualities must public speakers have that make it easy for them to keep their nerves under control? Mainly, they have confidence in themselves. And when you think about it, confidence is simply believing that you can do something. So it follows that if you believe that you can do something, then you can. It is as simple as that. You simply have to change your mindset.

When it comes to finances, you need only to change your way of thinking about money. Stop thinking of prosperity as something that is out of reach. We have already discussed that, when it comes to setting goals, specificity is very important. In financial matters, what you believe about your abilities is important too. If you change your habits, then the past really does mean nothing. It does not matter if you have never lived with the mindset of abundance; you can start changing that paradigm now and actually see how that will change your results for the better.

You do not have to walk all over others in order to be wealthy and to live in abundance. On the contrary, you must not walk over and take advantage of other people. If you change *yourself*, you will get new results, results that are completely different from those you got with your old paradigm. It is time to change your way of thinking about relationships, life, and, most certainly, success and prosperity. Remember that your harvest will be the result of the seeds you have sown.

There are people with bleak outlooks on life in general who bring that outlook to the workplace. This type of coworker can bring down the overall attitude of the entire office with constant complaining and grumbling. An unhappy person can easily make any environment toxic.

The last type of coworker that will certainly create a negative working environment is one who feels that he or she must always be in control. Coworkers such as these are the worst kind; they create an uncomfortable workplace because they do not follow Grandmother's Law. Instead, they find fault in everything, from big mistakes made by others to the way the office furniture is organized.

TOXICITY

But it is not only your coworkers that can make a workplace toxic. There are several other factors that contribute to this kind of negativity, such as workers being underpaid and the workplace having unfair practices. If, on the other hand, employees are treated with respect and not overworked or underpaid, the workplace can provide an amazing sense of accomplishment and have an atmosphere that promotes efficiency.

Second, and most important, a toxic work environment will lead you to believe that your choices are limited and that you cannot move on to a better situation. It can make you so unhappy that you could dread going to work every morning and be grumpy when you come home. This can lead to problems in other aspects of your life, such as your personal relationships. A toxic work environment can make you feel trapped, in a word.

For instance, a friend of mine was working at a job, every aspect of which she loved, except the environment, which was so toxic that it made her constantly unhappy. She had a boss who was unfair to other coworkers and who took credit for all the work that others had done. This boss was a difficult person to work for because she made it seem as if she worked a lot, when, in truth, she did nothing at all. She did not respect her employees and, in fact, would often steal their ideas and take all of the glory. My friend also noticed that she was stealing money from the company by paying herself out of monies reserved for other things.

However, my friend could not convince anyone of these facts and, day by day, she became more and more frustrated and miserable. Finally, after some convincing from her family, she decided to quit her job and find a new one. Eventually, she found a position that not only made her happy, but also paid more money than the previous one.

Not every work environment will be so toxic that it will be necessary to quit in order for you to be content and happy, but some work environments will be toxic. This may seem like an extreme situation, but if you think about it, we have all been in this position at one time or another. We have all had jobs that made us unhappy. What about the first job that you ever had? Was it one that you wanted to keep for the rest of your life? Probably not. So you worked hard or made changes until you were able to move on to another job that paid more, one that you also liked a great deal better.

WHY WE FEAR CHANGE

We fear change and we, in this particular situation, fear the loss of a career that is creating problems. No one would suggest that it is easy to simply change jobs or to find a job that makes you completely happy (or in which you can make enough money to live on). Oftentimes, however, it is the fear of change that is the very thing that makes us feel trapped, when, in reality, we can simply make the choice to change and then take actions that will allow us to reach that goal. If you let go of your fear and doubt, you will be amazed at the things that you can accomplish and the actions that you will take.

Let us consider the story of a man named Leon. Leon always wanted to move to the big city and start his own engineering business. Yet things kept getting in his way. He was unhappy where he worked, but he could not figure out how to change the situation. But Leon was only in his early twenties, and no one noticed that he remained at similar dead-end jobs for years.

As time went on, Leon's life stayed the same, while all of his friends moved on to bigger and better things. He still lived with his mother and family, and he was miserable. One day, I asked Leon why he did not just move to the city and live his dream. His response was, "I don't have a job there." I pointed out that it would be easy for him to get the kinds of jobs he currently had in the big city; they were only jobs, not anything leading to a career. He was afraid that he would not be able to find enough work to make it in the city; he was afraid that the novel he might produce would not be good enough for publication; he was afraid of a lot of things, and his fear effectively paralyzed him. He let his fear get the best of him, and he allowed his passions to be thrown to the wayside.

Leon never listened to me. He is thirty-eight years old now, he still lives with his mother, and he is still in the same type of dead-end job he has had for twenty years. He never moved to the city, never wrote a word of the novel he planned to write, and never became happy. What did Leon do wrong? Leon did not take action; he refused to change the paradigm he lived under and, therefore, he did not become successful. He never wrote the first page of his novel, and he never lived in his beloved city for even a week or a month.

Leon did not change these things because he was afraid to; he is still afraid at age thirty-eight. What do you think would have happened if he had taken my advice? Where would he be now if he had simply decided to change his life?

We fear change because we fear what we do not know. Again, it is a simple thing to change your life, but it is not always easy. Overcoming the fear of change can be a key element in your success rate and, therefore, in the kind of results you get. Just think of how many wonderful things we miss out on just because we are afraid of change. Certainly, this fear will affect our professional lives if we allow it to.

LETTING GO OF FEAR

Fear is often our enemy in many situations, but fear can be most damaging to us when it comes to our professional lives. Therefore, it is the ability to let go of fear and to overcome anxiety about making conscious choices and effecting change that will become important to us. Instead, we must focus on our passion and let go of our fears. But how do we do this?

Letting go of fear is not as easy as it may seem; if it were, then we would all be living and working to our potential; we would also all be happy. Remember that fear can paralyze us and make us think that remaining in a terrible situation is best because we fear change so much.

However, passion should, and will, overcome fear. There is an old saying that says, "Nothing ventured, nothing gained." This is what I mean: if we do not at least take small steps to overcome our fear, then we will gain nothing.

Overcoming fear means letting our passion take precedence over our anxiety. When deciding to let go of fear, we must remember how important our passion is to us. Whether our passion is to become an interior designer, a landscape architect, the CEO of a large company, or simply to have a job that allows us more time with our families, a conscious choice must be made to work toward these goals.

Simply put, you can decide exactly how you want your professional life to be. Decide what it is that you want, then take action toward making it a reality. Do not sit back and watch your life go by and your professional dreams slip away – make the choice to do it and take action. Otherwise, you will end up like Leon, never knowing what could have been if he had simply let go of his fears.

When we let go of fear and anxiety in following our professional dreams, we feel more fulfilled, and we gain a sense of accomplishment. Nothing in the world feels better than letting go, reaching a goal, and knowing that you did it all by yourself by simply deciding to do it and taking action. When you become the architect of your own destiny, you will understand how powerful an experience it is to change your life through changing paradigms that are not working for you.

EMBRACE YOUR POWERFUL NEW IDENTITY

Imagine that you could suddenly become a superhero. Superheroes are powerful, without fear, and they make a difference in their communities and the world. How would it feel to be able to make choices and to see the fruits of your labor? Just thinking of it will make you feel more powerful.

Okay, so maybe you're not a superhero, but you do have some of the same advantages that a superhero has, namely that you have the power to make conscious choices and to, then, act upon them. People rarely realize what making these kinds of choices mean for their professional lives. What you do for a living has to be something that is more meaningful to you than just a way to pay your bills; it has to be something that is important to you and makes your life full and significant. And the beauty of it is that you can have everything that you want by using my system.

We have discussed how releasing anxiety can lead to a better professional life, a new and rewarding career, and better results. But you must also learn to embrace the person that you are becoming; you must learn to embrace your powerful new identity. See, to make your own choices *is* a powerful thing; it makes you feel in control and confident. This is why making choices that are purposeful is so freeing.

THE BIG LIE

We have been convinced that we are supposed to be miserable in our professional lives. Media images tell us this, the work history of our parents tells us this, as does the behavior of our coworkers. Having learned this through experience, we have very little confidence in the idea that we can break free from our confining situations, from our cubicles, so to speak.

Our inner identities are formed by our subconscious mind from the way we are treated. There is no less power in the idea that we must work hard all of our lives, for very little money, doing something that we do not enjoy. Who made that a rule? Why is it so difficult to realize that we have control over our own lives? We do not have to live that way. What is the answer?

Make the decision and act upon it. If, for instance, you would like to become a photographer or an artist, then you can work toward that goal by embracing a new identity, one that is not fraught with misery. Instead, becoming a photographer, for example, might mean hanging on to your current job while you contact galleries, other venues, and people to display your photographs. What is certain, however, is that you will never make any positive changes in your life if you do nothing. Even making a mistake is sometimes better than inaction; inaction and fear paralyze us and make us feel small and weak, but action, on the other hand, will make us feel strong.

It is the biggest lie to think that we are trapped in a situation, especially a professional one, simply because we have been taught that it is somehow wrong to want something more out of life. Why should you remain in a job that you are not happy with? Because someone else has said that anything else is out of the question? Or because it seems out of reach, and it might be silly to dream about your perfect job?

Yet it is not silly to think this at all; it is not ridiculous. You can change your life, and you can do it by simply making the choice, but it involves a change in your mindset. If you cannot even imagine yourself in a different workplace or doing something different professionally, then change will most certainly never happen. However, if you can change your mindset, then you will be better able to take action.

THE MAZE

Imagine your current professional life as if you have been living in a world where nothing is what it seems. It is as if you have been trapped at the center of what you believe to be a maze. Every turn you make seems to lead to another dead end, and every wall seems to get higher as you go along.

Now, imagine that you only thought that those walls existed. What would happen if someone could show you that you have always known the way out of the maze, that, in fact, the maze never existed at all? It was the construction of some evil magician or, in this case, an image that is the product of your environment. What if you discovered that you had all the power over your situation? Wouldn't it feel good to know how powerful your new identity is?

If you want to change your life and escape from the maze of the professional paradigms that you have been trapped in, you simply have to make the choice to do so. Using my system, you can accomplish things that you never dreamed possible. You have the power within you to make any decision that you choose to make and to make your life better by taking action.

TAKING ACTION IN YOUR PROFESSIONAL LIFE

A friend of mine once told me that all you have to do, professionally speaking, is understand that you must choose what you want your life to be. Do you want to make more money? Would you like to be able to travel more? Is job satisfaction important to you? You must determine, first, what you would like your new life to be. The next step, once you have figured that out, is to take action in order to get what you want. You may be asking yourself, what steps can I take in order to really make my results better and to have more success in my professional life?

Much of this depends upon which career path you choose. It is possible that you already have ideas for just such an opportunity to occur. For many careers, some kind of training or education is involved, and you may feel that you do not have the time or the experience to enter into that kind of commitment. But this is simply your fear overtaking you and telling you what you may have thought all of your life – that passion is worthless and that it cannot ever be fulfilled. I know that this is not true.

There are, for instance, many people who go back to school after being in the workforce for several years or decades. Some of them are told that they must be crazy to go after another career at a "non-traditional" age, but they do not listen to this kind of negative talk. Instead, they know that if they allow their passion to take over, instead of their fears, that their reward will be greater in the end. And all the people who told them they were crazy will then be jealous; they might even want such results for themselves.

All non-traditional students made the choice to leave a career that was not working for them. They also took action by registering for courses and showing up for every class. Of course, they were a bit afraid in the beginning, but they took the chance and it paid off. They made the conscious choice to try to do something different with their lives and to do whatever it took for that to happen. These are the kinds of choices we need to make every day. We need to choose abundance.

One of the things that makes us happy, both professionally and otherwise, is knowing that what we do makes a difference, either in other people's lives or in our own. This is meaningful even if what makes the difference is that we feel more fulfilled, have more time, or have any number of things that are important to us. Professionally speaking, altering your mindset and making meaningful choices that will lead you to prosperity and happiness in the workplace will lead you to change your life.

CHAPTER EIGHT: Effecting Change in Your Everyday Life

There is no chance, no destiny, no fate that can circumvent or hinder or control the firm resolve of a determined soul.

—Ella Wheeler Wilcox

THE GRATITUDE ROCK

So far, we have covered the topics on how to effect change in your personal relationships, financial situation, and professional life. Now, we turn our attention to effecting change in your everyday life. Here we will discuss how to make the choices that will affect the specific as well as the general aspects of your life.

Each day you wake up and choose your attitude for the day. Most of us believe, especially at the beginning of the day, that we do not choose our attitude. Instead, we believe that our moods are something that happens *to* us, not something that comes from us. However, it is here that the ideas of personal responsibility and mindset work together – we choose our attitudes and actions, no one else. Our environment can only affect us as much as we allow it to, and each moment, we choose how we will greet the day: either with a positive mindset or a negative one.

Often, we need a little something to remind us to remain positive and to stick to our goals. People use different kinds of reminders: a piece of string on a finger, a note on the refrigerator door, or an inspirational quote taped to the bathroom mirror. The mechanism I use is something quite different, and I call it the "Gratitude Rock." My Gratitude Rock, a small rock from my favourite beach, helps to remind me to make conscious choices; it also reminds me of the importance of choosing a positive attitude. I keep it where I can easily see it or feel it and allow it to remind me of what I am striving for.

A Gratitude Rock can be anything that reminds you of what your goals are and that you should be grateful for everything that you have every day of your life. You will thus be able to begin each day with this reminder, and it will allow you to keep a positive attitude throughout the day. This will, in turn, make anything negative that happens to you a little bit easier to deal with. Find your own object to remind you of what you are grateful for, and keep it near so that you may begin each morning with a fresh, grateful, and positive feeling.

After using my Gratitude Rock for a while, feeling positive became a habit; it was integrated into my new self – a more positive and happy self. I can really feel the difference when I use my Gratitude Rock every day because I notice that I am much more able to keep a positive attitude, and this makes my results better throughout the day. It becomes second nature to remain grateful and to retain my upbeat viewpoint; it becomes part of my new and powerful identity.

People who keep a positive mindset, who feed their subconscious minds with only good and healthy things, are, in general, going to be more successful. No one wants to be around someone who is constantly negative. This creates the very situation we have already discussed: a toxic environment. Everyone knows someone like this, someone who is not interested in anything but negativity. This type of person will never be interested in solutions or in any new ideas; they are, instead, only interested in criticizing others. If you suggest any seemingly good idea to them, they immediately find fault with it. But where does this get anyone?

CATCH-22

Remember the old Joseph Heller book entitled *Catch-22*? Heller coined this oft used term; it refers to any situation that seems to have no way out. A catch-22 is kind of a "damned if you do, damned if you don't" state of affairs. Such circumstances keep quite a few people from understanding that there is a solution to their problem; they believe that they are caught up in a catch-22.

Once, I heard of a situation much like this in a university setting. The university had a good, well-respected graduate program, to which they wanted to add a doctoral program. The only problem was that the university could not fund a doctoral program until the graduate program had enough students in it. However, many people thought that the problem with finding enough students, on average, to graduate from the master's program was that there was no PhD program offered. Students generally don't want to do their graduate work twice in order to eventually earn a PhD; the proposition seemed preposterous.

It seemed that this was a catch-22. There were not enough students to create a PhD program because there was no PhD program. Yet it was my friend's attitude and mindset that was truly disturbing to me. Her argument against instituting a new PhD program was that it would not be competitive because it would be new. My answer was that if you do not begin a program, it will most certainly never be a good or competitive one; it will never have the chance to be so because it will have never existed.

I believe that this is how many of us live our lives. We worry so much about our finances, our personal relationships, and our jobs that we have very little else to think about. This leads to a poor attitude and the paradigm of scarcity. People who are discouraged are usually unable to deal with their problems rationally because they tend to panic and worry their way into more problems. I prefer to be more solution-oriented; I prefer to take a chance on something, rather than wondering what might have happened otherwise.

Living with the mindset of abundance and making conscious choices, which is really what the power of statistics is all about, is something that we can all do. But it takes action on our part. It means taking chances that we otherwise would not take. The end result will be a richer, more fulfilled life, yet many of us remain trapped in old mindsets.

OLD TOYS

Have you ever been in a discussion with someone, in which everything somehow revolves around what the participants had done in the past? One of my friends calls it the "Barbie argument," because she had a childhood pal with whom every discussion turned into "...and you broke my Barbie doll when we were six." The Barbie doll had no relevance to the discussion at hand, but it was brought up anyway, out of spite.

The past is very much like a series of old toys that we sometimes allow to haunt us. Past mistakes can't be overcome, and they are even harder to forget. This contributes to the mindset of scarcity and makes us afraid to try anything new. We live with the worry that someone will bring up that old broken toy and show it to us.

Keeping the image of our past mistakes in our mind begins to affect our subconscious and robs us of our confidence. When you were a child, the world was your oyster. You probably had a good idea about what kind of work you wanted to do when you grew up, and no one could have told you any differently.

We need to recapture the joy that we once had as children when everything seemed possible, because, really, that's a part of changing our lives so that we may live in abundance. We forget, sometimes, that we have the ability to change our situation, and this is what makes us feel trapped.

You can absolutely have all the things that you want in life, from living in abundance to making the money that you desire, to having lasting and meaningful relationships. The important thing to remember is that you have the power to choose it; that power is always at your disposal.

Many things in life distract us from our goals. However, worrying and living from the mindset of scarcity is the main thing that will prevent us from reaching our full potential. When you were a child, everything seemed possible; now that you are an adult, even more things are possible.

I cannot stress enough how much of your life and what happens to you in it depends upon your mindset and the choices you make.

Statistics can clue you in about what to avoid, but it is really the choices you make that will effect change. From the smaller choices to the bigger ones, you have the ability to make them consciously in order to reach your goals. You simply have to act upon them.

That is the problem that many of us have – the action portion. We understand what it means to have the potential to do great things, but to never act upon it. How many people do you know who have never reached their dreams? The dream could be something as simple as a desire to watch a particular sports team play, or it could be something as big as wanting to be the CEO of one's own company. But how many of these people have ever tried, really tried, to take action toward that goal?

AUDITION MARY

There is an old adage relating to the theatre that expresses similar sentiments. If you want to become an actor, you must simply never give up. The world of Hollywood and Broadway is a hard world to succeed in. Many successful actors have argued that all it takes is a strong desire and a powerful passion to become the working actor that you want to be.

This was certainly the situation with Mary Martin. Most of us remember Mary Martin as the petite blond who is famous for her roles in *Peter Pan*, *The Sound of Music*, and *South Pacific* on Broadway. However, it wasn't easy for Mary when she began her acting career.

Mary Martin's got the nickname "Audition Mary" because she showed up to every single audition she could. It became a joke because she went to more auditions than anyone else, but still never landed an important role. Yet she never gave up her dream of becoming a star and singing musical theatre on Broadway. She eventually got there with passion, determination, and action. Had she never gone to any auditions, she would have never received any important roles. There were even a couple of seemingly unlucky incidents, such as a car crash, that eventually led her to taking voice lessons and accidentally singing in front of Oscar Hammerstein II.

Ms Martin could have seen all of these things as obstacles, but they were not. They were "unlucky" incidents that came her way but were overcome by hard work and perseverance. She made her own choices and then acted upon them. Determination will lead to many positive things if you simply stick to the goal at hand. "Audition Mary" understood this completely, and she lived an abundant and happy professional and personal life because of it.

SUPERHEROES

Anything is possible once you use the power of making your own choices. What if Superman had never chosen to use his powers to fight evil? Or, better yet, what if Mother Teresa had simply worried about the world's poor but did nothing to help them? My program should inspire you to take some chances, to take some action, and to make your dreams a reality.

If there is something that you really want to do, then you should simply do it. If you work from the mindset of prosperity and abundance and make the purposeful decision to take action, then everything else will fall into place. If I want to go to a conference that is important to me, then I simply plan to go to the conference. I do not worry about money because I know that I have a goal in mind. If I keep that goal in mind, I can take action toward it by either saving money or getting someone to sponsor me. I do not worry about money because I know that it will come, and because I have made a choice to work toward my goal of attending the conference. The details will work themselves out later. Have you ever heard the phrase "Don't sweat the small stuff – and it's all small stuff?" This saying fits perfectly here – it is all small stuff. Worrying about your life will never get you anywhere; however, taking action will.

When this program becomes a habit, you will begin to notice radical changes in your life. In the beginning, the changes might not be as noticeable as you planned, but it will come. You will soon begin to understand just how much the paradigms you have been living with have kept you from living your dreams, and you will be grateful for the insight that others may not have.

Truly being free is having choices, and it is the greatest tragedy when we keep ourselves from succeeding and having a fruitful life. No one else can help you with this struggle; it is something that you must change from within. Only you can choose to make choices for yourself that will better your life.

People who are truly free to make their own decisions and to slough off the mindsets that are not working for them are happy. They are happy because they know that they control their own destiny, and that no matter what comes their way in life, they will be able to handle it, because they have the security of knowing that they can make a difference in their own lives. Don't wait for someone else to make decisions for you; make them for yourself. When you reach the goal that you are striving for, when you reach the top of your mountain, then you will understand how it was all worth the journey.

THE BUCKET LIST

The Bucket List, a movie starring Jack Nicholson and Morgan Freeman, is about the very subject we are addressing here. The story begins with two men who are both diagnosed with terminal diseases. They know that they might only have a short time to live, so they come up with a "bucket list" of things they wish they'd done before they "kick the bucket." At first, it is just a fantasy, a joke between two friends, a list that contains mere dreams, not reality.

Jack Nicholson's character, Edward Cole, decides that it is not enough to simply make a list of the things that they want to do; he suggests that they actually go out and *do* them. Carter Chambers (Morgan Freeman's character) is a bit afraid of this new venture and is torn between realizing his life-long dreams and spending time with his family before dying. In the end, Cole convinces Chambers to take a chance and to live his dreams. Cole argues that when you want something, you have to go out and get it. And in this situation, there isn't much time to fret over the possible consequences. There is only time and room for passion and determination.

Together, Cole and Chambers see the Egyptian pyramids, visit the Great Wall of China, skydive, and race some of the fastest cars around. They live their list and proceed to cross off nearly every single item before Chambers passes away. Cole and Chambers took action. They knew that there were things that they wanted to accomplish before "kicking the bucket," and they accomplished them. It was as simple as that. Why was it simple? Because they took action. Carter and Chambers made the conscious decision not to allow anything to get in the way of realizing their goals. They made a plan and simply proceeded to act upon it. The result is amazing, and the experiences and friendship enrich both their lives.

Cole and Chambers were in the same situation that we are in every day; they were simply more motivated. If you knew that you might pass away in six months or a year, you might live your life quite differently. Because you might never have an opportunity again, you might be more inclined to do things that you never thought you could do, or try things that you have always wanted to try.

If we live our lives like this every day, we would all be better off. There would be no regrets, and nothing in the world would seem impossible to accomplish. Therefore, in a real, everyday sense, creating goals and then taking action toward completing them is the way that everyone should live. Making those conscious choices, as Cole and Chambers did by "living the list," makes all the difference.

Every day, we choose what we will do. We choose our attitude for the day, whether it be negative or positive, and we choose how our environment will affect us. In addition, we choose what we will do about it. If you allow your goals to remain scribbles on forgotten scraps of notepaper, then you will have gained nothing. On the other hand, if you let go of all your fears and simply make deliberate choices about what you want and how to get there, you will be rewarded a thousand times over.

We can learn much from Cole and Chambers, and I suppose this is why *The Bucket List* is such a popular film. All of us want to believe that living in total freedom that allows for choice and action is something that is out of reach, a subject for Hollywood. However, your life will be as free as Cole's and Chambers's if you simply believe it is so.

Why should you believe that it is so? How can I prove to you that this is not a fairy tale? Because I know it to be true in my own life; I have lived this way and prospered because of it. You cannot convince me that the choices we make do not affect our everyday lives. Every movement we make affects some part of us and of our existence. When I wake up in the morning and I am in a bad mood, I can simply change that. I can decide not to be in a bad mood any longer. Sure, I use my Gratitude Rock to remind me of these things, but it is not my Gratitude Rock that does the work; it is me who does the work. I am the one who decides how my day begins and what will happen after that. Another way for me to get an extra good start at the beginning of the day is to think of my three-year-old son, Bob, who lives in Thailand with his mother. You can only hold one thought in your mind at a time; choose the good ones. Of course, I cannot control all the circumstances, but I can control how I react to my environment.

THE BUTTERFLY EFFECT

You many have heard of the "butterfly effect," but not truly understood what it meant. This theoretical concept proposes that a butterfly flapping its wings could cause any number of atmospheric changes. Generally, the idea is that any small change has the potential to change much larger things, like the butterfly causing changes in atmospheric conditions.

Certainly this is true in our own lives. One small change can change everything around us. Making one choice instead of another is similar to a butterfly flapping its wings. We may not see immediate results, but we have taken the small steps necessary to make large changes.

Even making one choice differently than before can change your life forever and in a positive way. Think about how many unconscious decisions you make each and every day. There are so many things that we do not think about that could affect our lives in larger ways.

For instance, when I am running late for an engagement or for work, I always think to myself, "I am going to think positively about being late today. Perhaps if I had left early, I would have gotten into a huge traffic accident and been seriously injured. Or perhaps I would not meet someone today that I am meant to meet, who will become important later in my life." The idea that everything happens for a reason is not new. However, when you really look at that mindset, it works quite well with my system.

If one change can change everything, both large and small, then imagine what immense control you have over your life right now, without even knowing it. You can effect change by making conscious decisions, decisions that will allow you to reach the goals you so passionately want to reach. Small changes will, like the butterfly's wings, change the rest of your life.

DESTINY OR CONSCIOUS CHOICE

A friend of mine, Maria, frequently lectures me on this very idea. She sees everything in life as an interconnected web, each piece affecting the others in ways that we can only visualize once they have already occurred. We never know, she tells me, if a seemingly meaningless choice might lead to an important result.

The example she uses most often is the story of how she met her husband. See if you can follow her line of reasoning. When she was seventeen, she had to decide whether to go to one university or the other. One university was several hours away from her home, family, friends, and her then high-school sweetheart. The other university was only one hour away from her family, and it was the same university that all but one of her friends would attend.

The education she would receive at the university farthest away would benefit her most, but she still had a hard time explaining this to her friends and to her sweetheart. When she decided to attend the better university, she felt confident that it was the best place for her to be. She was right. At that university, she made new friends who had similar interests and mindsets, and she flourished both personally and academically. She never regretted her decision, although she knew that her life might have turned out very differently had she attended the university where most of her friends were educated.

Many years later, after she had gone on to have a successful career, she visited her old university. She thanked them for having prepared her so well for her career, and she reconnected with many of her university friends who were still living in the area. On a subsequent visit, one of those friends introduced Maria to her future husband. In all the time that they had lived in that same small university town and attended the same very small school, they had never met.

Maria believes that everything happens for a reason, because she believes that if she had not chosen to go to a university that was better for her, a number of things in her life would not be the same. She might have married her high-school sweetheart and had no career at all. Or she might have ended up in a completely different career. Certainly, if she had not made that choice, she would have never known the one mutual friend that she and her husband had in common.

It just goes to show you that one decision, even one that seems inconsequential at the time, can change the direction of your life. Why not make conscious decisions in order to create a *better* life? Why choose to live in misery when the control that you seek is in your power. It always has been so; you simply need to listen to your instincts and passion and allow them to guide you. You must never allow the past to define you, however. That is dangerous business, just as it was for Orpheus in the following tale.

ORPHEUS AND EURYDICE: LEAVE THE PAST BEHIND

One well-known Greek myth, also a love story, is an appropriate example of the idea of having control over our choices. Orpheus and Eurydice were married and very much in love when the King of the Underworld steals Eurydice for himself. Orpheus travels to the Underworld and manages, through the use of his cunning and his lyre, to persuade the King to give Eurydice back to him. The King accedes to Orpheus, but with one condition: Orpheus cannot look back at Eurydice while he leads her out of the Underworld; if he does, Eurydice will be doomed to remain there forever. When they are not far from safety, Orpheus cannot help himself and chooses to look back, even though it means losing Eurydice for all time.

The story of Orpheus and Eurydice tells us, first, that we all have a choice, just as Orpheus had a choice. If he had kept his eyes on the goal of getting his wife back to safety, then he would have accomplished what he had set out to do and saved her. But he chose to look back and to destroy any chance of success.

Second, this story can teach us about how important it is not to look too closely at past events. Your future is now, and you can do anything with it that you desire if you simply believe in yourself and in your choices. Becoming the person that we wish to be is very simple, but it does not always happen in ways that we expect. When you live from the perspective of freedom and choice and realize how much control you really have, you will be astounded by how powerful that feeling truly is.

PASSING ON JOY

Another wonderful thing that happens when you have a positive attitude is that you begin to realize that it is infectious. Have you ever noticed that when you spend a lot of time around negative people that you become involved in their negativity? The people you surround yourself with will affect how you feel about yourself and about life in general; on the other hand, what you do affects others as well. If you are positive, then you will notice how much that attitude affects others and makes them positive too. Negativity spreads like a virus, but positive energy is like the rays of the sun, touching all who are within its reach.

Once you have used my system to take control of your life and of your choices by changing your mindset and subconscious thought processes, you will notice that your life as well as the lives of those around you will be full of abundant joy.

"FREE HUGS"

Not so long ago a man calling himself Juan Mann (One Man) walked out into a city center in Australia carrying a sign that said "FREE HUGS." This man had lived in Europe for a few years, but by the time he got back to

Sydney, most of his friends and family had moved on, and he felt that he needed human contact of some kind. So he decided to offer free hugs to any pedestrian who was interested.

Although most people at first looked at him as if he were crazy, an older woman was the first to take him up on his offer. As it turns out, the day that Juan was there, her dog had just passed away. It was also the anniversary of her only daughter's death. She told Juan all of this and gave him a huge hug; they both needed it.

Many people joined Juan, and they gave out hugs to anyone who took their offer. Sydney city officials tried to ban Juan and his group of huggers for fear of lawsuits, but 10,000 people signed a petition to allow Juan and the others to continue getting out their message of love.

Juan Mann is not the only person who has developed this system of giving free hugs. There is even a place in lower Manhattan in New York City, at Union Square, where people gather on the last Saturday afternoon of every month to share hugs. Most recently, the band The Skinny Puppies wrote a song and produced a video about Juan entitled "All the Same." The Skinny Puppies became interested in Juan when their lead singer gave him a hug. They became friends.

You can still catch that video on YouTube. Juan has made several appearances on *The Oprah Winfrey Show* and continues his work today. Perhaps you will begin a similar "FREE HUGS" mission in your city.

Juan Mann reminds us how little we really pay attention to the people around us. If we do pay attention to some folks, it is out of suspicion: we are afraid of strangers, or we are no longer as trusting as when we were children. We have been taught that it is wise not to speak to strangers, especially odd strangers such as Juan who hold up signs that say "FREE HUGS." Why is this so?

Certainly it is not a bad idea to be cautious, and no one would suggest that you walk into a dark alley, or allow someone to pick your pocket, or rob you, if you can help it. However, not everyone in the world is that way. We have learned that fear and distrust are the only things that seem real to us.

This is not the case. The people who have hugged Juan Mann have no reason to be afraid. He simply wanted to pass on to them some of the joy that he had in his heart. He didn't want money or fame or recognition. Juan wanted, and still wants, everyone to get a sense of community that people used to have. Juan wants people to have contact with and help one another in small and large ways. We all know that we need human contact, love, affection, and consideration.

Without these things, our lives would be empty. We would feel very alone and distraught. So why is it that we have gotten into this habit of being suspicious of one another? Where has this fear of one another come from? It comes from our environment and from society. People are taught to "look out for number one" and to walk all over and step on people to get to the top.

Is this really how we all want to live? Is it possible that there is another way, one that allows us to remain wise and cautious, but to let go of our fear and suspicion of others? I think there is a way to do this. You must simply decide that you will and then take action. Throw aside your fears and create a new life for yourself.

This is exactly what Juan Mann did. He needed to restore his faith in humanity and in himself. Instead of fearing the world and worrying about whether or not he would be laughed at, he decided to take a chance by standing at the street corner, holding a sign that read "FREE HUGS." Some people must have thought that he was very silly. Yet there were others who did not think that Juan was silly or stupid or crazy. There were some that needed to remember what living in a community was all about. Some people were happy to express and pass on their joy to Juan; they were happy to see him making a difference in other people's lives.

Our everyday lives can be difficult to get through. Sometimes each day seems to be a struggle. We have families to take care of and very little time to ourselves. Perhaps this is why so many of us get caught up in paradigms that are not working, and we do not even realize that we have choices. It's easy to become complacent and to think that things will never change. But they will change if you create change.

Be inspired to challenge yourself to use Grandmother's Law on others and on yourself as well. Think of yourself as the person Grandmother's Law is directed toward, and find something positive about yourself to tell yourself every day. When you wake up, during the day at work, and before you go to bed, remember to find at least one wonderful thing about yourself, and tell yourself about it. You could use Post-it notes, as I did with Mara, or you could simply think about it as the day goes on. Perhaps standing in a busy city square and holding a "FREE HUGS" sign is not your thing, but I am sure that there are many other things that you can do.

Tell yourself that you are in control of your life and that the past is no longer important. Remind yourself that you do not have to live with the mindset of scarcity, that you have the choice to live in abundance. It comes down to making choices, acting upon them, and being positive; it is a simple choice to make. In fact, I once read somewhere that if you are upset, you should make yourself smile. I do not mean that you should find something funny to read or a silly website to peruse; I mean, find something to literally make yourself smile. The act of smiling releases endorphins in your body that make you feel better and happier. This will lighten any bad mood and keep you positive, thereby creating a cycle of positive energy that will constantly encircle you with its quiet power.

CHAPTER NINE: What if the Odds Are Simply Against Me?

When you get into a tight place and everything is against you, till it seems as though you could not hold on a minute longer, never give up then, for that is just the place and time that the tide will turn.

—Harriet Beecher Stowe

THE ODDS

Thus far, we have discussed many things about the power of statistics. However, we have not covered the problematic issue of what should be done if following the program seems too difficult. What if the odds are simply against you?

If the odds seem to be against me, it is often a matter of perception. Some things that are seen as roadblocks by one person might be seen as stepping stones by another. When I look at situations in the past in which I thought the odds were against me, I now see that my perception was often dependent upon how I used statistics to interpret those situations. When I interpreted them one way, I saw them as obstacles to overcome, but when I took a longer look at them, I realized that they were simply challenges that could easily be overcome with a little hard work and perseverance.

At the time I applied for the Military Academy, for instance, the statistics said that the odds were against me. Only forty out of three hundred that applied would get in. However, I had preparation and hard work in my favor, and I succeeded.

When beginning this program, you may feel a little uneasy. After all, you are embarking on a wonderful journey, but it is to a place that you have never been. Shifting paradigms and mindsets takes some work, but it will soon pay off. The way to avoid being discouraged in many situations is to simply be well prepared. If there is a test that you must take, the best way to feel confident about doing well on that test is to study, right? Well, this is no different; being well prepared will make all the difference. If the odds seem to be against you, prepare for whatever the situation is and work hard toward your goal.

RUDY'S STORY: "IT'S ALWAYS TOO SOON TO QUIT"

Rudy Ruettiger was one of 14 children of a Catholic family living in Joliet, Illinois. His family had no money; his father had to work three jobs just to put food on the table. Rudy had little athletic ability and was often teased in school about his tendency to daydream. One of his teachers told him that the problem with dreamers is that they rarely take action. But Rudy was not what he seemed to be.

Rudy's dream was to attend Notre Dame University, and nearly everyone that he first shared this dream with thought that it would never be possible. Notre Dame is not an easy university to get into. The odds were against someone like Rudy, who did not have the build or the talent to get an athletic scholarship, nor the grades to get into a university of that caliber. So he temporarily gave up his dream and went to work in a factory. While working in that factory, a close friend of his died, and this event made him decide that he had better get his life together and start really working on his dreams.

Because he did not have the grades to get into Notre Dame, he decided to attend Holy Cross Junior College, which was right across the street from the Notre Dame campus. It was at Holy Cross that Rudy was first diagnosed with dyslexia. Finally, he understood why he'd been having so many problems in his classes. He worked hard and learned how to study all over again, this time aware of his learning disability. His eventual goal was to transfer into Notre Dame, but in order to reach his goal, he had to keep at least a B average at Holy Cross.

For years, Rudy struggled with dyslexia, but he worked hard and managed to maintain the average grade he needed. After being rejected three times, Notre Dame, in the fall of 1974, finally accepted him. There, he concentrated on academics, but by then he had another dream: to play football for the famous Fighting Irish, Notre Dame's team. But Rudy Ruettiger was only 5 feet, 6 inches tall and weighed only 165 pounds. He was not exactly a perfect fit for the football team.

However, a kind coach who encouraged walk-on players from campus allowed him to join the team, but to play only during practices. Rudy treated each and every practice session as if it were a real game. He struggled to reach his goals; it was not easy for him. After two years on the practice team, Rudy had one opportunity to run out of the tunnel at a real game, to proudly wear the Notre Dame uniform, and to sit on the bench next to his teammates. The coach would not allow it, but the other players thought that Rudy had earned the right to be there, and they fought hard to convince the coach, who eventually relented.

The day of the big game, Rudy was ready. Every single day he had set his goals and worked hard to accomplish them. It did not matter how high the price was or how tough it got; he had persevered. And being a part of the team that game day was proof that Rudy was ready for anything. He had the opportunity to make a key tackle during that game against Georgia Tech that made the difference in their winning the game, and he was carried off of the field in victory on the shoulders of his teammates.

The odds were against Rudy from the beginning. His family did not have money, he had a learning disability, and he had no athletic ability. Yet he still became a football player at a prestigious university, and he now travels the country as a successful public speaker.

What is important to remember about Rudy is that he did not let the odds affect his actions. He stopped being a simple dreamer from the time he began working toward his goals, his dreams. He took action. Here believed in himself and his dreams when others saw his dreams as impossibilities. There were mentors along the way that helped and encouraged him to live his dreams, and they inspired him, just as I hope his story will inspire you.

Rudy was able to change his mindset and reach his goals. It took years of hard work, yes, but he did it by making deliberate choices that were informed by the statistics involved. Even though he knew that his grades were not good enough for Notre Dame, he found a way: he went to Holy

Cross instead and brought those grades up. He knew that the odds were against him when he decided that he wanted to play football, but he found a way to do just that. Rudy was never discouraged, he never believed that he would not reach his goals, and he never ever gave up. He made a conscious choice to go to Notre Dame and also to play football there. The way that he got there was through simple perseverance and hard work.

Rudy's story is inspirational; in fact, there was even a film made about his life in the early 1990s. He still tells anyone who will listen that no matter what your background is, no matter what statistics are against you, you can still do anything you set your mind to. You simply have to take action instead of merely dreaming. Preparation is a key element; you must earn your dreams.

CHANGING YOUR LIFE

Changing your life is simple, but not easy. I cannot stress how important this point is. There are very simple solutions to all of your problems. There are certain simple steps that you must take, but for which you will need a strong will and a burning desire for success, much like Rudy.

Take a look at the things that you have done in the past that seemed difficult at the time. Many of them, in hindsight, will not appear to be quite so hard. This is because you took action and have accomplished those goals already. Think of anything difficult that you have accomplished; was it really that difficult?

Believing in yourself is also of great importance. If people do not believe in their dreams or, more specifically, in a project, then they will be doomed to fail before they even begin. This is where a positive attitude comes into play. It is certain that nothing will get accomplished if nothing is attempted. Do not pay attention to the statistics. The power of statistics lies in knowing that they do not affect you. You have the power to change your life on your own; there is no need to become discouraged by what others say. Was

Galileo Galilei discouraged when everyone around him said that the earth was not the center of the universe? No. Did Christopher Columbus get discouraged when he was told that he could not sail around the globe? No. And you should not get discouraged either during the process of changing your life. Persevere and you will succeed; you will find a way.

PAST FAILURES

If you're trying to achieve, there will be roadblocks. I've had them; everybody has had them. But obstacles don't have to stop you. If you run into a wall, don't turn around and give up. Figure out how to climb it, go through it, or work around it.

—Michael Jordan

Our past failures seem to follow us around like a black cloud. They make us believe that we are not as strong as we truly are, and they can affect every aspect of our lives. You may begin to think that all of your past failures are predictive of how you will fail in the future. But this is not the case. Simply because you have had failures in the past does not mean that you will fail in the future. Your history does not dictate what will happen to you now.

Scientists often comment that the reason that experimentation is so difficult is that you must repeat an experiment a number of times to be sure about its results. If they are the same the first 99 times, it is still possible that the hundredth result might be different. If well-regulated scientific experiments can end up with different results, even though the experiment has been performed many times over, then you too, in your life, could end up with different results than the ones you have had in the past.

The results you get today are only dependent upon the amount of hard work and action that you put in. If you are dedicated to creating circumstances that are of your own choosing, then you will get the results that you want. Through deliberate choices, one can be assured of the best results.

DESIRE AND PASSION

You may be wondering about the reasons behind your past failures. Perhaps you tried making changes or doing other things that did not work. The problem probably was a lack of burning desire, of not really wanting or believing enough in those changes or in that project. A burning desire is most important when attempting to reach your goals.

We have to truly want what we seek in order to succeed. If I go into something halfheartedly, the results will certainly not be as good were I to have gone into it wholeheartedly. Every great success story has an element of the underdog in it. When people say that the odds are against them and that the task at hand is impossible, I note that they have just described every great success story.

POST-IT

Don't be surprised if you have never heard of Spence Silver or Art Fry. Very few people in the world know or remember the names of the inventers of Post-it notes and products. Theirs is a success story that involves a struggle against the odds. They used their ingenuity and perseverance to invent and market one of the best-known products around – the Post-it note.

Spence Silver worked as a chemist for 3M Company, a business with plenty of resources. When he experimented with various substances as a chemist, he was accustomed to experiencing the bane and joy of all scientists – trial and error. You see, it was Silver's job to create a new type of glue, but as his experimentation continued, he realized that he had created a very different type of substance indeed. He had created a kind of anti-glue. The glue he invented would not permanently stick to anything. As a matter of fact, it didn't stick very well at all. Whatever Silver attempted to use this substance for, it would not stick properly to any surface, but would slip off.

Instead of becoming frustrated, Silver recognized that the polymer could probably be put to good use in some other way. Surely there was a way to utilize the substance for some other, as yet unknown, purpose. He began experimenting again, this time with the properties of the non-sticky glue he had created. At first, he tried using it on bulletin boards, but that didn't work.

Silver decided to ask his fellow employees at 3M if they had any ideas on how to employ this substance properly. In the beginning, no one had any ideas – no one, except Art Fry.

At the time, Art Fry was the choir director at his local church. He was having a problem during rehearsal one day that he couldn't remedy. The pieces of paper that he used to mark the pages in the hymnal would constantly fly away, and he would thus often lose his place in the book. Then Fry thought of Silver. He wondered whether the glue could be used on small pieces of paper to make them stay put on various surfaces.

Fry immediately brought the idea to Silver. Silver gave him a sample of the non-sticky glue substance, and Fry went to work, attempting to use it on pieces of paper. At first he tried using a large strip of the substance on a piece of paper. This did not work. He tried in on one of his hymnals, and it worked well until it was time to remove it. Then it stuck too well and nearly destroyed the pages of the book.

Silver and Fry did not give up, however. Fry tried adding a chemical primer to the adhesive that Silver invented and found that, this time, it worked. The "sticky" notes were now capable of being used and reused. They tested different types of models with their fellow coworkers until they determined that the small, colorful notepads were the way to market the invention.

3M was not behind this project in the beginning. It seemed that marketing and, especially, production costs would be difficult to handle. However, once they saw that the "sticky" notes could be used in offices to mark things and to send memos, they decided to take a chance. Now, Post-it notes are famous around the globe, and most offices could not do without that accident of an adhesive that Silver originally invented in his lab.

It's funny how perseverance and action can take your ideas to the next level. What was originally an accident in a chemist's lab has become a well-used and well-liked product, a common feature of modern life. Neither Silver nor Fry were willing to give up when they thought they had reached a roadblock. Instead, they continued to press on until their idea was fully developed and understood. What would have happened if Spence Silver had simply considered his strange polymer a fluke and an unusable accident? The world would never have known about the useful product he and Fry had created.

Post-it notes may never have come into being if Silver and Fry had not kept a positive attitude. What could be more positive than turning a mistake into a fortune? When you want something, you must make choices that will lead you to that goal, and then you must act upon those choices, just as Silver and Fry did. They did not allow anyone to crush their dreams or ideas. On the contrary, they continued to argue their point until 3M, and others finally saw the benefit of their product. Do you recognize the company name 3M without immediately thinking of Post-it notes?

NEW VENTURES

You should never think of your past mistakes or failures when beginning a new venture. If, in the past, you did not believe enough or work hard enough, it does not matter; what matters is what you do now. We have already discussed that changing your paradigm is difficult. Changing your life is no laughing matter, but it is simple to do so, almost too simple.

Many of us have a hard time believing that all it takes is belief, perseverance, and action to change our lives. Yet this is most certainly the case. History is chock full of stories of people who had to overcome great odds in order to get what they wanted out of life. When explorers set out to discover the New World, for example, very few people supported them. In fact, many naysayers said it was impossible to circumnavigate the globe because they believed that the earth was flat. People who believe that they can sail around the world, they argued, will simply find the end of the world and fall off.

It is that simple to make meaningful alterations in your life that can lead to happiness. Making the decision to change is half of the battle, and taking action is the other half. We cannot remain complacent in our lives when our lives are not working for us. And why should we, because we get discouraged? Everyone gets discouraged sometimes, but as long as we continue fighting the good fight, our lives *will* change and our dreams *will* come true.

Just because you have failed before does not mean that you will fail again. The past does not dictate what the future will be, and our past actions do not define what our future accomplishments will be. Yet, somehow, we have a hard time believing that we control our own destinies. There is nothing surprising in that; it is a very simple idea. You control your own feelings and those feelings govern your actions.

If you have a negative attitude, it is within your power to change it; it's not complicated. When you feel there is no hope for a change or that there is an impossible task in front of you, just believe in your powerful new identity and take action.

REMOVING OBSTACLES

The obstacles in your path to success may seem numerous, but this is simply an illusion. When you learn to use my program, you will find that all of the obstacles you once believed were in your way are suddenly gone.

This is because these obstacles are not real. It is your decision whether or not to see an obstacle in your path. In the case of Rudy, there were many things that seemed to be obstacles in his path, yet he managed to overcome every one of them to reach his goal. You can do the same. Any time that you reach an impasse in your life, just remember that the choices you make are what give you the power to overcome it.

We create many of the obstacles that block our progress. When we have the mindset that we are incapable of doing anything meaningful or good, we set up a roadblock for ourselves that is not really there. We have discussed in earlier chapters how a negative self-image and self-stereotyping can keep us from reaching our potential. It is the same case here. If you believe that you will never succeed at a particular task, then you are tying your hands behind your back. Believing or thinking that something cannot be done is the first hurdle that you must overcome. Anything is possible with your new and powerful identity. You must never forget that it is completely up to you.

Yet fear and doubt often creep into our minds. We think to ourselves, "Can I really do this? Will I make a fool out of myself if I try this new venture?" You must immediately put all of these fears and doubts to rest. They are the only things that can keep you from living in abundance.

NON-TRADITIONAL

I have several friends who have decided to return to university in the United States at an age that is not typical for a college student. Many of them are afraid that they will not be able to do well in their classes because they are too old, and many of them have not done schoolwork for a long period of time. But these are imaginary barriers that must be ignored or destroyed. They have nothing to fear, yet they have doubts about everything. They doubt that they will fit in; they doubt that the other students will respect them. They even wonder about their own intellectual abilities and how the course load will affect them.

I always tell them that these things are not obstacles at all. Whether or not they become friends with their younger classmates is completely irrelevant to reaching their goals. Respect will come during the course of the first year or so when the younger students begin to realize that their older, non-traditional counterparts are valuable resources of information and wisdom.

Accomplishing their goals will, of course, require a great deal of work for these non-traditional students. They face different challenges that the younger students will not understand. However, I have never seen anyone work quite so hard at their education than these friends of mine.

They do not play around in class. Turning in their work in a timely manner and spending time making it the best product they can is second nature to them because they have more experience in the real world. Most non-traditional students, in fact, turn out to be some of the most dedicated and hardworking students in universities. Why is this? It is because they know what is at stake, and because they are determined to attain their degree, having overcome many obstacles to be at the university.

Allowing fear and doubt to plague your mind will make your life nearly unbearable. Constantly worrying about what people will think of you or how you might change yourself to fit into someone else's vision is absolutely absurd. When we fear things and doubt ourselves, we are only setting ourselves up for failure. What else could be the result of fear and doubt? These factors can paralyze even the most brave and intelligent person from taking action toward their goals. They do not allow us to move forward.

Even if you are afraid, you must press on anyway. The funny thing about fear is that if you do not confront the source of your fear, your fear will grow. In our discussion earlier about the "golden mean," we found a description of courage that fit into the idea of the "golden mean" perfectly. Courage, you will remember, is being afraid but doing something anyway.

If we can remember this fact, then it will help us to face those fears and doubts. Just remember that you can handle anything that is thrown your way. This is not dependent upon what kind of person you are at the moment or whether or not you are perfect and consider yourself a courageous, strong person. You are a strong, courageous person anyway, without even having to be aware of that fact.

When you act like a strong and courageous person, you become one. Acting like a person who has no fears and doubts will make you a person who has no fears and doubts. Many people who seem courageous, strong, and successful to us would be quite humble if you were to compliment them upon their accomplishments. They would not even realize that they had done something important or amazing. What you expect of yourself is the eventual result that you will get.

If you expect that you will consistently fail, then you most likely will fail. Having that mindset will affect every aspect of your life and color your every choice. You will fail because your mindset believes that to be the only possible outcome.

On the other hand, if you believe yourself to be a strong, courageous, successful, and powerful person, then you will become that too. Because all of the choices you make on a daily basis will be from that mindset, those choices will lead you to success.

REMOVE THE OBSTACLES

If you want something, anything, you must work hard for it. Removing obstacles that prevent you from living in abundance requires absolute passion. You must want to reach your goal more than anything else in the world, and you must maintain an attitude of diligence in order to keep from becoming disheartened.

Remember when you first learned to do many of the things that you take for granted every day? For example, there was a time in your life when you did not know how to read, or to even tie your shoelaces. Yet you learned all of these things at an early age, and now do them with skill and ease as an adult. However, when you were first learning these mundane things, they seemed difficult to you. You did not learn to tie your shoes the first time you attempted it, so why should making changes in your life be any different?

After a short struggle, however, you learned all of those things, and now you probably do them without giving them a thought. They have become second nature to you, just as making conscious choices and acting on your goals will become once you have begun. Once you've started upon it, it is easy to continue along one path than to change to another.

The same thing happened to you when you became comfortable with the current paradigm. You didn't realize change was happening; it simply became a habit after a while. That is why it is so frightening to try something new and to effect change. We worry that things will not turn out the way we want them to; we would rather hang around, living in our old mindsets because they are familiar to us.

When we notice the roadblocks of our current paradigms, we become leery of change and fall back into old and unhealthy patterns. However, because it is easier to continue in one paradigm than to change it, it will also become easy for you to remain in your new paradigm once you have really made it a part of your life.

It will become effortless for you to think positively, live from the mindset of abundance, and to reach every single goal you set for yourself. This goal could be as simple as wanting to lose weight or as complicated as a desire to change career paths. Whatever your goals might be, you will find that your new paradigm will feel just as comfortable as your old one, and you will have the added benefit of getting what you want out of life.

GOOD PLANNING

Destiny is not a matter of chance, but a matter of choice; it is not a thing to be waited for, it is a thing to be achieved.

—Williams Jennings

This is where good planning and preparation comes in. Working hard should always be part of your master plan, but hard work alone will get you nowhere. You must work toward a goal, but not randomly and without purpose.

Let us imagine that you want to write the great American novel. You could not simply sit down and begin writing it. Writing all day long might produce several thousand words on the page, but the effort would be without purpose. With a novel, for instance, you would have to create characters that interact with one another and then come up with situations for them to interact in. It would be impossible for you to work without a distinct framework or plan.

So you would probably write an outline and make notes about the characters in your novel long before you sat down to the actual writing process. This is the attitude that you should have about everything in life. You must have a plan of action before you begin. If you plan and prepare well, then you will be ready for whatever opportunities present themselves to you. If you are not prepared, then all of the hard work and passion in the world will not make a difference; your goal will remain a dream if you do not persevere.

IT'S ALL WITHIN YOUR REACH

Absolutely anything that you wish to do is within your reach, but you have been programmed to believe that it is not. Everything we learn in school and from our environment is designed to teach us that dreams are silly, crazy ideas. This is because we fear change, because we fear the unknown. But what if the unknown is better than our current comfort zone? What if everything you dream of having is within your power to possess? You can.

The mind is a powerful tool, and it can factor greatly in our decision making. This is why belief is so important. Once we believe something is true, such as our self-image, it becomes true. There is nothing magical about this concept; it simply is. I am not suggesting that we can create things in our minds like purple elephants and make them appear before us. I simply mean that perception is reality. What we perceive to be true becomes true because we give it the power to exist.

If I have a negative self-image, then I will not believe myself to be a worthwhile person. I will react to every situation during my day as a person who is not worthwhile. I might pass up on promotions or personal relationships because I do not believe myself worthy of them. All of this, my actions in particular, wind up making my supposed unworthiness true. I become a person who is not worthwhile to others because I treat myself that way and thus invite others to follow my lead.

However, if I believe myself to be capable of handling anything that comes my way, to be a person of great strength and determination, then I will act like a person who possesses these qualities, eventually, therefore, becoming a person with these qualities. It is as simple as that. The ways in which we act, the choices we make, and our beliefs all work together to determine how successful our lives are and whether we live from the mindset of abundance or of scarcity.

The wonderful thing is that these things are under our control. We can choose to think differently about ourselves and, thereby, project that image to the world. People who are positive project a positive image and pass on the joy of positive feelings to those around them. People who are strong, passionate, and hard working will always be successful because they project that image to the world.

You have a vision of what you want your life to be like, and you have the power to realize that dream, down to every detail. Determine what your goal is, what kind of life you want, and then make choices that will create the changes that you want. Like Rudy, it won't matter to you anymore what obstacles are in your way, because you know they are imaginary. There is nothing standing between you and abundance except yourself. It is up to you to determine how badly you want something in this life, because if you really want it, you will find a way to get it. You will find a positive way, a way that does not involve immorality or cheating. The path you take will lead you to simply remove all the obstacles you face from your mind and to take action.

You will find that most of the details will work themselves out because you will have attracted other like-minded individuals to yourself without even trying. Have a purpose, a vision, and a plan, and you will indeed beat the odds and find yourself with a better return. With determination and perseverance, you will find that you can accomplish anything that you set out to do. But you must begin by letting go of any fear of change that you may have; you must keep your mind open to new possibilities. For it is only when our hearts and minds are open and we maintain a positive attitude that we recognize true opportunity when it comes our way. And if we are prepared for that opportunity, if we have worked hard and been absolutely passionate, then we will be able to accept it when it comes.

If you have a burning desire for something, you will make it happen. You will find a way to make it happen that is not only positive, but also beneficial to you and to others around you. This is the heart of the power that you possess, the power of statistics.

The very fact that you now hold this book in your hands and are reading it is an example of my theory. I started the process of being a LifeSuccess consultant some five years ago when I first met Bob Proctor, an internationally renowned motivational speaker and leader in the field of personal development. His program, "The New Lead the Field," awakened my interest about the mind and how it works. My interest grew stronger and culminated with my certification as a LifeSuccess consultant in the spring of 2008. One evening, I spoke with Gerry Robert, who has written a successful book and now helps others publish their books, for about an hour and then decided to write my book. At that time, I did not know what topic I would write about, but the decision to write a book was made. The process started with planning the book and then writing it, and now you are able to read it.

CHAPTER TEN: Getting the Best Return – Consistently and Often

The difference between the impossible and the possible lies in a man's determination.

—Tommy Lasorda

LOSING YOUR MARBLES

Not so long ago, a coworker sent one of those "forwards" to my email account. It was about Saturday mornings and how relaxing and beautiful they can be. The email was also the story of how a woman listening to a radio program one Saturday morning was touched by a story about a man and a small ritual he had involving marbles.

It seems that this man had a theory about gratitude and passion. He told the radio host about his "thousand marbles." At first, no one knew where this theory was going, but he soon explained that there was a ritual that he participated in each and every Saturday that involved 1,000 marbles. You see, this man had heard somewhere that the average person lives about seventy-five years. He did a little multiplication and discovered that, at seventy-five, a person would have lived through about 3,900 Saturdays. At age fifty-five, he figured that he had about 1,000 Saturdays left.

So he decided to go to the toy store and buy 1,000 marbles, one for each remaining Saturday. He then placed them in a clear container in his workshop, and each Saturday, he threw one marble away to represent the passing of another week. This wasn't just some odd way to pass time, some morbid means by which to mark the passing days. His marble ritual allowed him to see how many Saturdays he had left and to appreciate each and every one of them. It made him remember to cherish every day and not to waste a single moment. In fact, he even made the comment that if he gets rid of all 1,000 marbles, he will know that he has truly been lucky, because he will have had more Saturdays than the average person.

A single marble can be just as unique and wondrous as a Saturday afternoon. One Saturday is not at all like another; each one is special in its own right and should be appreciated as such. Marbles are each so different, just like days. Taking a marble out for each passing week reminded him of the fact that Saturdays, and every day, are things to be enjoyed and not wasted. We can learn a lot from this man and his marbles.

If we were to treat each and every day as if it were the last, then we would be happier. This would truly be living in abundance. Having a tool, such as marbles or my "Gratitude Rock," is important, in that it helps us to remain on track and to remember how powerful we are and how important the time we have is.

START NOW

In order to get a better result consistently and often, you must make important changes in your life. However, many of us are tempted to simply continue to think about the changes that we would like to make in our situation and never to do anything about it. I have already mentioned that action is important. This is because the sooner you begin to use this program, the faster you will get the result. Only action gives results, thinking does *not*.

Why wait any longer to change the paradigms that are not working for you? You should begin now so that you will get results more quickly. If you sit around and wait, you run the risk of never changing your life. Acting now to implement this program is the best thing to do, and it will make you feel good just to start to make changes in your life that you know will end in positive results.

Make a test for yourself. Start with just one small thing and see what happens; say something positive to a person you like or are close to, for example. Check out how you feel after doing this, and watch the reaction of the other person. I guarantee you that you will feel good and receive a positive response. Then you can go on and test it further, and after a while it will become a new and valuable part of you.

We all fear change and this fear prevents us from reaching our full potential as human beings. It stops us from acting on our dreams and from having the freedom to make our own deliberate decisions. Fear creates the belief that action is not possible. We become paralyzed.

Inaction is part of many people's mindset. They have a propensity toward inaction, and it becomes difficult to get out of that comfort zone. For instance, as a child, were you the type who did your homework as soon as you got home so that it was out of the way, or the type who did it at the last minute?

Why do we fear action as much as we do? It is because we fear change so much that it is foreign to us to think that we can do anything to change our circumstances. However, this is not entirely accurate. It is not our circumstances alone that we strive to change, but also how we react to our environment that makes a difference. Yet we are afraid to act because we fear bad results. On the contrary, if we do not act, we will not lose anything, but we will also not gain anything.

We all tend to procrastinate more than we should, but this is not something to procrastinate about. Procrastination is not a useful tool, and it will not give us better results. When we procrastinate, there is a chance that, in the end, we will have to do a shoddy job in order to get the work done. Also, when we procrastinate, we can never relax or remain positive because our responsibilities are constantly hanging over our heads. It is always better to complete a task as soon as possible. Unexpected events in our lives serve to distract us from our goals. Because life itself can often get in the way of our dreams, we must always be prepared for any unknown elements.

If, for instance, there is a project that I must complete and I believe that I have ample time to do it, there is always the possibility that something else will come up and prevent me from its completion. Therefore, I prefer to get things done as quickly as possible so that I will have plenty of time and energy to do them if something unexpected pops up, or if I simply wish to ride my motorbike on a fine day instead of worrying about work!

You must act now if you want results in the future; procrastination will not allow you to progress. Everything in life, especially major changes, is made better when you immediately act upon your decisions. This is not to say that you should not take time to consider your decisions; they must be deliberate and lead to your eventual destination. The process is important, but action is paramount.

ACT QUICKLY

Once you have made the decision to act, you must act quickly. Otherwise, it is too easy to fall into complacency and not act at all. Acting quickly is always a good habit to get into with anything that you do. Have you ever noticed that the things in your life, from phone calls to large projects, always get done more quickly and better when you work on them immediately after you get the idea? After you have made a conscious choice, you should act upon it immediately.

This will be the time during which you are most passionate and excited about the choice you have made; you will have top-of-mind awareness about it, and you will not have to wonder why you are so excited in the first place. If you lose that passion, then it will be difficult to get it back. Your attention and patience for the task at hand will wane, and you may never get much done at all.

Acting quickly will also set you on your path more quickly. As you progress, you will find that your path will become easier to traverse. You will gain more confidence, and a sense of accomplishment will cover you with positive feelings. Never underestimate the positive feelings and confidence that come from completing a difficult task.

I once heard someone say that if it is not difficult to complete a task, then it is not worth completing in the first place. Here, the main problems lie in both beginning the task and in remaining true to it until the end. A clear vision and purpose are important here, because the promise of a reward will help you to avoid complacency and keep your passion fresh. I have for many years experienced that doing the things that I am good at is much easier than doing the things that I am not that good at. It is the same with practical versus theoretical things. My subconscious mind makes it easier to do these things because it knows that the results will be good. When it comes to doing new things, it is much worse. I fear the result and thus procrastinate.

SIR ISAAC NEWTON

This idea can be best explained by Sir Isaac Newton's First Law of Motion. You may not remember the name and the term from an old science class years ago, but you will understand the concept. Newton's First Law of Motion states that a body that is in motion will always tend to stay in motion unless acted upon by an outside force. It also states that a body that is at rest will remain at rest until it is acted upon by an outside force.

In other words, if something is in motion, it has to be acted upon by some other force to make it stop. Conversely, if something is not in motion, an outside force must make it move. Here, you are the outside force; you are the one with the power to change your mindset and shift paradigms. Once in motion, you cannot be stopped unless you choose to stop.

You are the force that will allow a paradigm shift in your life. Change occurs through the action of an outside force; you have to make it happen. Remaining at rest is what most of us do in our lives. We allow things to happen to us instead of taking charge and making our own choices that will lead us to getting what we really want out of life.

But you don't have to remain in the old paradigm. You can change your outlook and what happens to you. Then you will become a body that remains in motion instead of a body at rest. Action will become second nature to you, and you will suddenly find that it is just as easy to remain in your new paradigm as it was to be trapped in the old one. In fact, it will be much easier because this paradigm will be one of your own construction, fashioned out of deliberate choices and actions that are yours alone.

A CLEAR VISION: MAPPING THINGS OUT

The power of statistics will help you to get a better return because your motion and action will not be without purpose. Statistics will be the guide of your actions; it will not spell doom for your future. Statistics does not hold power over you; instead, it gives you the ability to get the results that you want.

Having a clear purpose and vision is of the utmost importance, and it is similar to being well prepared. However, instead of preparation alone, what will guide you is a clear vision and purpose: a definite goal to work toward. If you do not know where the road you are taking will lead you, then it is more likely that you will get lost and become a victim to one of the many pitfalls we have already discussed.

This is why you need a map. Your map will assist you in being certain about where the road leads. We have already discussed how to avoid the dangerous areas of your journey; let's now focus on the eventual end of this process.

First and foremost, you must decide what kind of life you want. What is success to you? Will you have reached it when you earn a certain amount of money per year, or when you have the house or apartment of your dreams? Are your aspirations in life tied to people more than to things? Is it your personal relationships in which you want abundance?

You can have all these things and more, but you must know what you want from the beginning. If you have no specific goal, then reaching it will seem too difficult because you won't be following a map of any kind. Instead, you will be floundering around, making aimless and senseless choices in an attempt to get, well, somewhere. You will not get very far if you do not know where you want to end up. This would be like going to a travel agent and telling him that you want a ticket. He cannot help you unless you tell him where, when, and for how long you want to travel.

What will your map be like? What is your eventual goal? Decide where your map will lead to and then set out upon the path immediately. I cannot stress enough how having a clear purpose and a clear vision will benefit you when beginning this program.

If we have a definite goal in mind, it will seem easier to reach. We are able to see how far we have come and how far we need to go. You can break this goal down into subgoals. This process will make it easier for you to see and understand that it is possible to reach the goal. If, for instance, you have a major goal that you must accomplish by the end of the year, it is helpful to break this goal down. If you can ask yourself, "What do I have to do today and tomorrow to reach this goal?" then the major goal becomes easier to manage. This will give you a sense of accomplishment and encourage you to continue to strive. Michael Jordan once said, "I've missed more than 9,000 shots in my career. I've lost almost 300 games. Twenty-six times I've been trusted to take the game-winning shot and missed. I've failed over and over again in my life. And that is why I succeed." Jordan's perspective is one we can all learn from. Success is the fruit of failure.

When I am, for example, separated for long periods of time from the ones I love, I find the first days to be the most difficult. After a certain amount of time goes by, I stop thinking about how long it will be until I see them again. Once I reach a certain point, I can say to myself, "They are coming soon; I can hold out until I see them again." Being able to track that period of time in my mind is helpful, because once I am "over the hump," so to speak, I am more excited about my goal than I am discouraged about the past.

When setting your goals, you should aim high. Why not? If you make choices that will lead you to them, and you act upon them and you work hard, it will not matter how far off your goal seems; you will be able to reach it. Nothing is impossible once you believe in your own powerful new identity.

Having a goal is important for all of these reasons. It gives us something to work for, it directs our actions, and it keeps us from becoming discouraged because we can see the progress that we have made. If we have no idea of where we are going, then it is difficult to see how far we have come. However, if you never begin your journey in the first place, then you will most certainly never reach your destination. The sooner you begin upon the path set forth by your map, the more quickly you will arrive at your destination.

CONSISTENT RESULTS

In order to get better and consistent results, you must fully concentrate on the task at hand. Once you have successfully caused a shift away from your old paradigm, you must remain passionate about your vision. The clear choices that you will make will allow you to continue on the correct path. But you must have the necessary determination in order to keep yourself going. Becoming discouraged is not an option; remember that any obstacles that seem to be blocking your way to abundance are merely imaginary and have no power over you.

In other words, for your results to be consistent, you must also be consistent. You will not find this hard to do once you get started, because soon it will become like second nature to you. Getting better results, like a positive attitude, catches on. It will begin to affect all aspects of your life in unexpectedly wonderful ways.

We have stressed the importance of your choices being deliberate and conscious, but what does that mean? It means that for consistency and results, the choices that you make must be directly related to reaching your goal or goals. Each decision should be tailored to shifting paradigms from a negative mindset to a positive one.

Remember the idea of the maze? Think of your decisions as the twists and turns that will lead you out of the labyrinth. One by one, your decisions will lead you to a better result, to the exit of the maze, and into a brilliant world of abundance and light. Suddenly, the "maze" that you have been living in will seem more like a wide-open field. Remember that part of the paradigm most of us live under includes a fear of the unknown. It is this fear that you must overcome in order to release yourself from a maze of misery. There really is nothing to fear and no obstacle in your way that you cannot overcome or go around. Rudy Reutigger taught us that.

Once you have begun, you will understand how easy it will be for you to be consistent. It will be as if many things that you thought stood in your way have mysteriously disappeared because you are able to see things so much more clearly.

THE MATRIX

This film is perfect for describing what happens when you create a shift in your paradigms. Most of us are familiar with the story: a young man, Neo, finds that his world is not exactly what it seems. He discovers that there are people and things within his world that make very little sense. But he does not know how to prove this, even to himself.

Along comes Morpheus, a veteran of this kind of thinking. Morpheus explains to Neo that the world he lives in is simply an illusion. Everything that Neo believes exists does not really exist; it is all a construct of his mind. He then offers Neo a way to see it for himself; he offers him a choice. Neo can take one pill and remain ignorant of his situation, or he can take another pill and understand what Morpheus calls "the matrix."

Neo chooses the path of understanding and finds that all human beings have become batteries for machines that now rule the earth. The images pumped into their minds to make them believe that their lives are normal are called, as we soon find out, "the matrix," a grid created by machines in the minds of human beings. Even déjà vu is explained away as "glitches" in the matrix.

It is a strange tale, but it mirrors how a paradigm works and how determined one has to be in order to escape it. When Neo decides to truly explore the matrix by taking a leap and trusting Morpheus, he finds that, because they are merely false images, he can control the matrix and, therefore, his own life.

The matrix is much like a paradigm. We do not often know that we are trapped within a particular mindset, but we are, in reality, in control over everything we do. We can control the matrix; the matrix does not control us.

If we are, therefore, consistent and confident about this fact, then our results, too, will be consistent. Although once you have really begun to implement this program, you will find that it is not difficult at all to remain consistent. Your new paradigm will be so comfortable that you will never want to return to any other way of living.

It is all up to you to determine what your results will be. The more energy and determination you put into reaching your goals, the better your results will be. The more consistent you are, the more consistent your results will be. The power is in your hands. It is a great responsibility, but not difficult to accomplish. Once you realize how powerful you are, you will want to remain consistent because you will want your results to remain consistent.

In general I am more consistent when I receive positive feelings and feedback from my work. When I get better results, I feel better about myself and my level of confidence rises. I then pass this on to everyone around me, and it, therefore, makes my entire life better than it was before. This joy and positive energy will then come back to me and a circle will be created, where everyone feeds off of one another's joy instead of misery. Others will be inspired to break free from their confining paradigms, and they, too, will begin to see how powerful true freedom is.

Consistency will come if your efforts are consistent. If you truly desire for something to happen, then you will find a way to make it happen. As we have discussed, however, it must be done through your belief in your own power, not by walking all over others. The only true way to get a better result consistently and often is to depend upon yourself to make the necessary shifts in mindset that will make your life full of abundance. A clear purpose and vision are part of that plan, but these come from within you.

Begin effecting change in your life now so that you do not have to wait any longer to live abundantly. There is no reason in the world for you to continue on the path you are on and not to choose a new path. The change may not be immediately apparent, but you will begin to notice subtle changes and then larger changes. Let go now of any remaining anxiety you may have about changing your life. Let all the tension and worry fade away. You will begin soon to live a life of true abundance over which you have complete control.

CHAPTER ELEVEN: What Will Getting the Best Return Do for My Life?

Formulate and stamp indelibly on your mind a picture of yourself as succeeding. Hold this picture tenaciously. Never permit it to fade. Your mind will seek to develop the picture... Do not build up obstacles in your imagination.

—Norman Vincent Peale

CHANGING DIRECTION

Using this program will gradually change your life for the better. Beginning with the new choices that you will make and the actions that you will take, you will see a brighter and more positive life for your future. When you look at statistics of any kind, you will know that you no longer fall into those old categories. Instead, you will have the peace and strength of will that will come from your powerful new identity.

No longer will you have to be part of the masses of people who have not discovered this for themselves. You will have the tools that you need, through this program, to change your life forever and for the better. Just think of how much richer your life will become once you decide to take action and begin to make the necessary changes.

We all know that making such radical changes can be frightening. We have already discussed how fear of change is a huge problem for many people. They fear making changes because they are either afraid of doing something new or of what will become of them once they leave their comfort zone. Some people are even afraid of change because of their failed attempts at change in the past. Remember, failures can be viewed as stepping stones to success. People of action must, by the laws of nature, experience failures. But this does not stop them from trying.

As I have said, all of these fears are simply not relevant. You will find that once you begin, this program will be the best thing that has ever happened to your life, your finances, and your personal relationships. The roadblocks that you see now will not matter to you once you are fully immersed in this program. They will no longer even exist because the new perspective that this program will give you will change that. It will change how you see everything in your world and how you deal with any issues that you might have.

This program is more than a simple way to get the things that you want out of life; it is a way of life. It will change the way you see your finances, your professional life, and your relationships. The power that this program will give you is the power to control the things that happen to you. You will no longer have to remain in an inescapable situation of any kind. Instead, you will understand, without even having to think about it, what you must do in order to make your life what you want it to be.

In fact, the larger purpose of this program is to give you the knowledge that you can get through anything and create your own life. Everyone wants more control over his or her life, and this program is the best way to get to that place. Once you have begun to implement these ideas in your own life, you will gradually understand the full implications that living according to this program suggests. Getting the best return will change your life forever, and for the good.

STATISTICS: WHAT WE CAN LEARN

Using statistics to effect the outcome of every facet of your life is one of the best ways to ensure that you lead a happy, healthy, and productive life. You are not held captive by statistics; instead, they give you the power to avoid negative consequences in any given portion of your life. They are like road signs pointing in the opposite direction from where you want to go. Statistics are lessons learned and signs of the things that we should avoid; they are cautionary tales full of data that is only significant because of its power to warn us.

Your path lies in a different direction than the one that statistics point to. The things that you have learned about conscious decision making will be informed by the relevant statistics on any subject. Once you have fully immersed yourself in this program, you will understand that your decisions will be conscious because you will now know how to avoid the negative consequences that statistics point to.

For example, doctors prescribe drugs to their patients, even though they know about the potential negative consequences. They prescribe it anyway because the benefits outweigh the risks, such as the side effects. This is the way in which this program works when it comes to statistics. We can foresee what will happen and avoid the negative things happening to us.

Armed with that knowledge, we can identify both the real and imaginary obstacles that lie in the way of our success. Gaining knowledge is truly half of fighting any battle. When you are knowledgeable about the statistics, you can better avoid bad things happening to you. In fact, you have the opportunity to redirect your actions away from these things. You are absolutely in control of all of your actions, and the sooner you begin to realize that the power to change your life lies within your grasp, the sooner you will begin to live in abundance.

THE NEW, POWERFUL YOU

If I were dropped out of a plane into the ocean and told the nearest land was a thousand miles away, I'd still swim. And I'd despise the one who gave up.

—Abraham Maslow

Embrace your new and powerful identity completely and with excitement. Knowing that you are free to create your own life is to know ultimate freedom. We spend far too much time blaming ourselves for things that we cannot control, when the truth is that we can simply change it. If you have made mistakes in the past, forget them.

Those mistakes have nothing to do with what will happen to you in the future. Begin now to make changes instead of feeling guilty about things that have happened in the past. We are often too hard on ourselves about the mistakes we have made. This program will allow you to understand how to let go of all of that fear and worry, to trade in those thoughts of living in scarcity instead of in abundance.

When you begin to have the self-confidence that comes with using this program, you will never want to return to your old paradigms. The feeling will be so amazing, the freedom so full of hope, that you will soon forget the way you used to live; you will never even be able to imagine living that way again. Many of us could use a bit of this kind of self-appreciation. We have been conditioned to continually criticize ourselves and to think that we can do little to change our lives. This is why this program is so important. It will allow you to understand just how powerful you truly are and to tap into that power through conscious decisions that will lead you to your goals.

When we talk about paradigms, it can be confusing to many of us. Paradigms are the systems of belief that we live under, and they have power as well. This is why it is best for you to make a shift in those paradigms, to move away from any system of belief and mindset that is making your life something that you do not want it to be. If you want to change something, then you simply have to decide to do it and then take action, led by those decisions.

If you want to change your career, you need only to do it. If you want your personal relationships to be better, then you simply have to take action to that effect. You need never feel powerless or alone in this enterprise. Sure, you may feel a little out of sorts at first, but this is just part of adjusting to a new and more powerful life.

THE CAVE

One of the most well-studied example of this kind of paradigm shift is found in Plato's ancient work, *The Republic*. It is known as the "Allegory of the Cave," and it symbolizes what happens when people attempt to become enlightened by realizing that they can change their circumstances.

The story begins with a group of people who are chained to the inside of a dark cave. They can see nothing but what is in front of them, because their chains bind them in such a way that they can only look forward. What they see are shadows cast on a wall by a fire that is just behind and above them.

They do not know that what they are seeing are merely shadows; they believe the shadows are real objects instead of shadows of objects. Because they have never known any other existence other than the current one in the cave, they have no way of knowing the difference between reality and shadows.

One day, one of the people trapped in the cave decides to escape, and when he attempts to leave, he notices that his chains do not actually hold him in this prison. It has been his ignorance of any other way to live that has kept him there. He runs past the guards that are stationed at the mouth of the cave and escapes into the bright world above.

At first, the man thinks that he has made a huge mistake by leaving the cave. Everything around him is foreign and the world is too well lit for him to see very easily, as his eyes have only known the darkness of the cave. He has trouble distinguishing light from shadow and becomes afraid.

Gradually, however, his eyes become accustomed to the world above, and he slowly begins to look around him. He notices that what he now sees as a tree, for example, is very different than what he saw as a tree within the cave. Now he understands that everything on the walls of the cave were shadows and not reality; they were poor approximations of a real tree or lake or anything else.

The man is amazed at what he has found. Once his eyes adjust to the outside world, he discovers an entirely new reality waiting for him there. He understands more than the difference between shadows and reality; he also understands how powerful this makes him. The knowledge of the difference between what is real and what he always thought was real sets him free. He is free from the cave and from his own misconceptions about his life.

The cave represents the paradigm that many of us currently live in. It represents a place in which we have truly trapped ourselves, a place in which we have forgotten our own power over making changes in our lives. The cave is a place of false reality, but it allows us to understand that many obstacles in our lives are not real; they are merely imaginary, merely shadows.

We have lived as if we were in a cave, trapped by the paradigms others have imposed on us. Escape is easy: decide that you will do so and act upon it. You need not live in the shadows of the cave; you can leave the darkness, move into the light, understand your powerful nature, and truly accept your new identity.

This kind of freedom is difficult to understand because many of us are accustomed to the dark, and we never ask why we are there or whether or not there is another way to live. We have so little faith in ourselves that we cannot imagine that our lives could be good. Our belief that our lives are not good will lead to our lives continuing to be bad. What we believe becomes reality.

Have you ever heard of a self-fulfilling prophecy? A self-fulfilling prophecy happens when we have so little belief in ourselves or in a situation that what we do not want to happen does happen because we are afraid it will. If I believe that I will not do well when I interview with someone for a new job, then I will not do well. Believing that I am an unlucky or unhappy person will make me an unlucky or unhappy person. This is not because my thoughts contain some kind of magic, but because what I believe affects every thought that I have and every action that I take.

If I believe myself to be a horrible person, or if I do not love myself completely, these self-perceptions will be transferred to the world around me. These beliefs may be subconscious, but they are powerful nonetheless.

On the other hand, if I follow this program and allow myself to believe that I am powerful, then I will be so. I will be able to accomplish anything that I want to because I will believe from the beginning that I can. Confidence and passion are important because they keep us from getting discouraged. They remind us that we are wondrous beings who have complete control over our own destiny.

Why should we remain in the shadows when we could live in the light? Why should we not get what we want out of life? Is it because we have been taught not to be too ambitious, and to accept our lot in life with dignity and resolve? Or is it because we have been told one too many times that we will never amount to anything more than what we are now?

Every success story in the world involves a struggle against the grain. It always includes a little inventiveness, hard work, and perseverance. To be successful, you must take responsibility for yourself and for the things that happen from this point on and into the future. Do not drive yourself crazy thinking about the things that you should have done; look to the future and act now to make sure that the future will be exactly what you want it to be.

CONTROL

This program, the power of statistics, can help you, in short, to get better control over your life. Making a difference in your life is simple, but it is not always easy; we have discussed this at length. When you have control over your life, everything seems so much easier. It is not a simple matter of perception, but one that involves perceptions. Having a positive mindset that is connected to your unconscious thoughts is imperative. A positive mindset will allow you to take control of your life more easily because you will no longer see everything in a negative way. Instead, you will see everything as having purpose and meaning, a positive meaning. You will easily overcome each new obstacle because you will no longer see it as such.

Think of everything that you have learned about this program and about gaining a new and powerful identity. Now think of someone in your life, and there is always someone who has not been successful or lived in abundance. Usually, there is an element of negativity in such people; they are certain that nothing will ever go their way, and nothing ever does. We call this being unlucky, but is it, really? Or is it just that person's perception of what has happened?

THE LEMONADE

Perception and belief are enormously important tools in trying to change your life. Have you heard the expression, "When life gives you lemons, you make lemonade"? This is the same basic idea. If you see the lemons only in a negative light, then you will never understand their part in creating a positive consequence – the lemonade. The lemonade is what we are all looking for. Call it abundance, or call it a positive outlook; it's all still lemonade. It is representative of all the good things that we all want – a house, a family, a high-paying job doing something we like to do.

How we see the lemonade and the lemons that make them is what really makes the difference. Perception is everything here, and a shift in paradigms is simply a shift in the way that you perceive the world. If you perceive the world as a place in which you have no control, then you will make that a reality by seeing everything in that light.

However, you have the power to overcome those perceptions and to make deliberate choices that will lead you to any goal that you can dream up. Again, statistics can guide us by showing us what we ought not to do, by showing us how to avoid the pitfalls, or lemons, that life might throw at us occasionally. But through your own choices that are not influenced by any negativity or fear, you will become a person with a better outlook and a richer, fuller, and abundant life.

When you make choices that are free of influence from your old paradigms, the situation creates a change in your very way of being; it creates a mindset in which you can flourish and be full of gratitude. If you are truly prepared, mentally, for the opportunities that you expect will come your way, they will arrive. More importantly, you will recognize them when they come along; you will no longer fear change because you will have already accepted it.

Once you realize that you have control over your life and that the choices you make are entirely up to you, then you will find that you are like a new person, a person who is in complete control and is, therefore, happy. Old problems that might have bothered you a great deal in the past will no longer seem so large and burdensome. Instead, you will take everything that happens to you in stride because you will be safe in the knowledge that you live under a different paradigm than those around you – a positive and goal-oriented paradigm over which you have total control.

YOUR LIFE

Go where your passion leads you, dream any dream that you wish. But you must do your part as well. You must determine what your goals are and then have the passion and the determination, like Rudy, to make them come true. Even when all the odds seem to be against you, you must remember that this is an illusion. The odds are never actually against you; this program will give you confidence to take that leap into the unknown, to abandon your fears, and to begin truly living your life.

It is your choices that are the most important. What you choose and whether or not you act upon those choices is up to you. If you follow this program, you will find that it will change your life, not only for today or for the moment, but forever. Your whole frame of mind will be different and, therefore, your life will be different.

No longer will you be a slave to your habits, your paradigms. The choices you make will come directly from your desires instead of habit or some other random, meaningless source. You will actually be making choices yourself, not simply thinking that you are choosing, when, all along, your paradigm was choosing for you.

Remember that your choices will affect both the small and the large aspects of your life. Every aspect of your life will change for the better once you begin to act of your own accord, without any outside influences. This is what every person dreams of – to be able to effect change in every section of his or her life without any outside influence. That is the real dream – freedom from habits that keep us from living the way we would like to live.

You can beat the odds by taking care of your body, as well as your habits. If you choose to eat healthy foods, to drink clean water, and to have a positive outlook, you are certain to live a healthier, longer life. It is true that stress, worry, and unhealthy habits are not good for you; these things can cause a tremendous amount of health and other problems that will make your life difficult. If, instead, you allow yourself the freedom and the respect to act upon what you choose and thus be more relaxed and ready to take on the world, then you will live longer.

You may consider these suggestions for your life too simple to work. You may be wondering how merely changing the way you think about your life can actually change it. But this program is that simple. If you do not want to become a statistic, you do not have to. By the same token, if you do not want the job you have or the relationship you have or the level of self-esteem you have, then you should change them.

There is an old Chinese proverb that says that if you find something wrong, it is your responsibility to fix it. Or, as my friend one time put it: "If you need to speak with your boss, go to the mirror… he is always available." The proverb and saying apply here because the only one who knows whether or not there are things in your life that are not working for you is *you*. No one else, not your family, your boss, your coworkers, or your friends, can tell you whether or not you are stuck in a negative paradigm. Only you can recognize it; only you can change what happens to you. Only you can decide whether or not you are satisfied with all aspects of your life.

If you are not satisfied, however, then this is the best program available to you. Part of the reason that I choose to share this message is because it has worked for me. I live from the mindset of abundance each and every day. I make sure that my choices are well thought out before I act upon them because each choice I make is tailored to fit my goals.

I am constantly aware of having a positive attitude and of passing it on to others. It has truly made a difference in my life – in all parts of my life. I find that my personal relationships are more rewarding and that I have developed a stronger and more solid relationship with my family. All of this is so because I am aware that each choice I make affects all of us.

Beyond that, I have found my finances are no longer a concern for me because I learned to stop trying to limit my potential to make money. There is high demand for my services and I am not easily replaceable. In my professional life, I have found true satisfaction and learned to never fear change or new elements in my career.

NO LIMITS

Your potential is limitless. The more quickly you begin to make choices that will lead you to your goals, the better you will understand this. There is really no end to the success that you can have once you let go of the old habits that are holding you back. Change your mindset and change your life.

Nothing can prevent you from wielding this new-found power that you have always possessed. And you have always possessed this strength, this confidence. However, up until now you may have been unaware that so much of what happens in your life is truly up to you. You may have been convinced, until you discovered this program, that your future was completely out of your control. Perhaps you were in a rut, a slave to your habits and subconscious thoughts. Let this be the last moment of your life in which you feel powerless. Let my program teach you the way to a better and happier life. Allow this program to help you to help yourself and to change your life forever.

Why wait? There is a life of abundance and success that awaits you, if you decide you want it. I know how much of a difference the program has made in my life, and I want to share that with you. So do not wait any longer: live the life that you deserve to live; begin using this program today; and soon you will understand how powerful you will begin to feel.

All of your choices will be meaningful, your goals worthwhile, and you will find yourself becoming a healthier, more positive, and more confident person. People will be drawn to your obvious success and abundance and will want to know your secret. Tell them that your secret is the power of statistics. Why shouldn't you live the way you choose to live? Why shouldn't you be successful when it is you who have the control? You can change your life now by beginning this program today. I know how using this system has created significant positive changes in my life; I want the same for you. So begin using the power of statistics today to take action and to live a life of abundance.

NOTES

NOTES

NOTES

NOTES

NOTES

NOTES

NOTES

LaVergne, TN USA
15 January 2010
170106LV00002B/2/P

9 781599 302751